Introducing Sedimentology

Other Titles in this Series:

Introducing Astronomy (2014)

Introducing Geology ~ A Guide to the World of Rocks (Second Edition 2010)

Introducing Geomorphology (2012)

Introducing Meteorology ~ A Guide to the Weather (2012)

Introducing Mineralogy (2015)

Introducing Natural Resources (forthcoming 2015)

Introducing Oceanography (2012)

Introducing Palaeontology ~ A Guide to Ancient Life (2010)

Introducing Tectonics, Rock Structures and Mountain Belts (2012)

Introducing the Planets and their Moons (2014)

Introducing Volcanology ~ A Guide to Hot Rocks (2011)

For further details of these and other Dunedin Earth and Environmental Sciences titles see www.dunedinacademicpress.co.uk

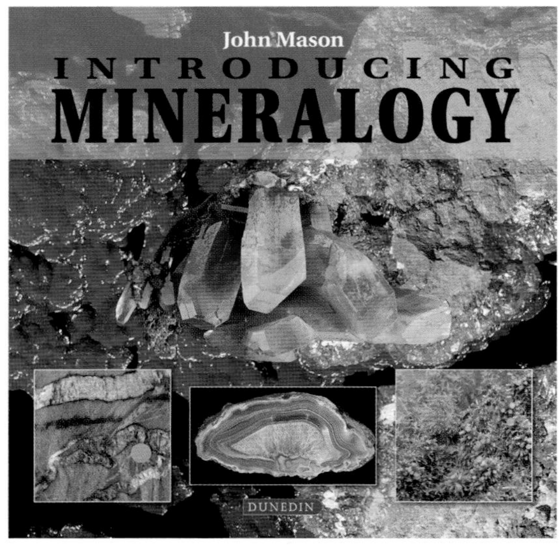

INTRODUCING SEDIMENTOLOGY

Stuart Jones

EDINBURGH ◆ LONDON

Published by
Dunedin Academic Press Ltd
Hudson House
8 Albany Street
Edinburgh EH1 3QB

London office:
352 Cromwell Tower
Barbican
London EC2Y 8NB

www.dunedinacademicpress.co.uk

ISBNs
9781780460178 (Paperback)
9781780465319 (ePub)
9781780465326 (Kindle)

British Library Cataloguing in Publication data
A catalogue record for this book is available from the British Library

Typeset by Makar Publishing Production, Edinburgh, Scotland
Printed in Poland by Hussar Books

Contents

Acknowledgements

I have received considerable support and advice from many colleagues in Durham during the preparation of this book with particular thanks to David Harper for assistance with fossils and exceptional preservation, Howard Armstrong who is always willing to offer an opinion and Maurice Tucker who is inspirational for everything sedimentological. Enormous thanks are due to Chris Orton for all of his hard work in drawing many of the figures in this book. A large chunk of the book was written while I was on research leave at the University of Waikato, Hamilton, NZ and I particularly thank Peter Kamp for his hospitality and teaching me about the last 65 millions years of New Zealand's amazing geological evolution.

This book would not have been possible without the continued unrelenting support of Nichola and my two little field assistants, Adam and Luke.

List of illustrations and tables

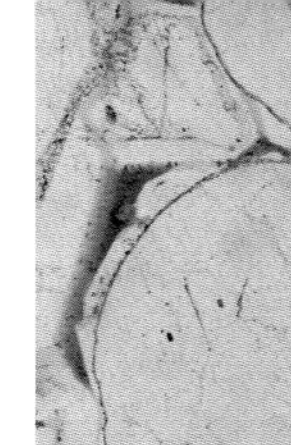

Preface

For at least the last 3.8 billion years of Earth's history sediments and sedimentary rocks have accumulated providing an archive of past surface environments and the evolution of life. The Earth's dynamic behaviour over geological time, mainly driven by plate-tectonics, has allowed a vast array of sedimentary rocks to form and it is these and the associated processes that eroded, transported and deposited the sediments that provides Earth with a uniqueness in the Solar System. This book is intended to provide some insight into sedimentology and the study of 'soft rocks' in different environmental settings and identify why sediments are so important.

Sediments and sedimentary rocks decorate much of the surface of the Earth and form many of the wonders of the world, such as the Grand Canyon, the Great Barrier Reef and even the summit of Mount Everest. It is perhaps this variety that drew me to sedimentology as a truly fascinating sub-discipline of Earth sciences. As a sedimentologist with a diverse interest in all sedimentary rocks and associated processes I have tried to represent this in the book from using my own field observations, experimental work and their applications. Perhaps some of the greatest advances in sedimentology have occurred over the last decade with improved knowledge of deep sea environments through sea-bed imaging, the interaction of climate and sea level as recorded by sediments and the recent images from NASA's Curosity Rover providing a remarkable insight to the sedimentological record of Mars. However, it is likely that the greatest challenge still lies ahead for sedimentology with the search for new ore minerals hosted in sedimentary rocks, the need for improved recovery of current and new oil and gas reserves and the growth in natural gas production from sedimentary shale formations.

It is hoped readers will find this book a useful guide to sedimentology and beyond all, use it as a guide to find out more about the planet we live on.

Dr Stuart J Jones

Note: all terms initially highlighted in **bold** are defined in the Glossary at the end of the book.

1 What is sedimentology?

In this book **sedimentology** is introduced, a branch of geology concerned with the nature and origin of **sediments**. It includes the weathering of rocks to produce sediment, their dispersal and deposition in different environments from terrestrial to the deepest parts of the oceans. The sediments are transformed through time by chemical, physical and biological processes to form sedimentary rocks.

The term 'sedimentology' has only been widely used for the last few decades to better encompass the improved understanding of the processes that create sediments and the dynamic environments where sediments accumulate. In many ways, sedimentary rocks advocate the Uniformitarian principle *the present is the key to the past* better than anything else in geology, and sedimentary rocks have much to tell us about their history, and in turn the history of the earth. Since some 70% of the rocks on the Earth's surface are sedimentary in origin, and sediments are of great economic importance, there is a very good chance that we encounter a **sedimentary rock** or an associated sedimentary process at some point in our daily lives.

Sedimentologists strive to understand the Earth's history from studies of sedimentary rocks through fieldwork, field observations, laboratory and theoretical studies. Sediments and sedimentary rocks yield important information about palaeoclimatology and palaeogeography, and contain the record of life on Earth, preserved as fossils. The fossils provide the basis for **stratigraphy** as a direct relationship between rocks and time. Sedimentologists need to cross many different disciplines and sub-disciplines of geology and ocean sciences to fully understand and appreciate the sedimentary rock record.

From grain to basin

All sedimentary rocks are composed of sediment that is varied in its origin, composition, shape, and size of grains. Solid particles or grains may be sourced from the erosion of other rocks, precipitated directly from solution by chemical processes, and include whole or broken shells created by once living organisms. This sediment provides a wealth of information about the way in which the material was carried and deposited. To create any package of sediment, transportation of grains by water, air, ice and/or gravity is required. Biological growth or chemical precipitation of sediments can also take place *in situ*. Transport and processes operating where sediment accumulates result in **sorting** (Fig. 1.1). Sediments are characterized as poorly sorted when the grains are of differing sizes, and well sorted when grains are about the same size. During

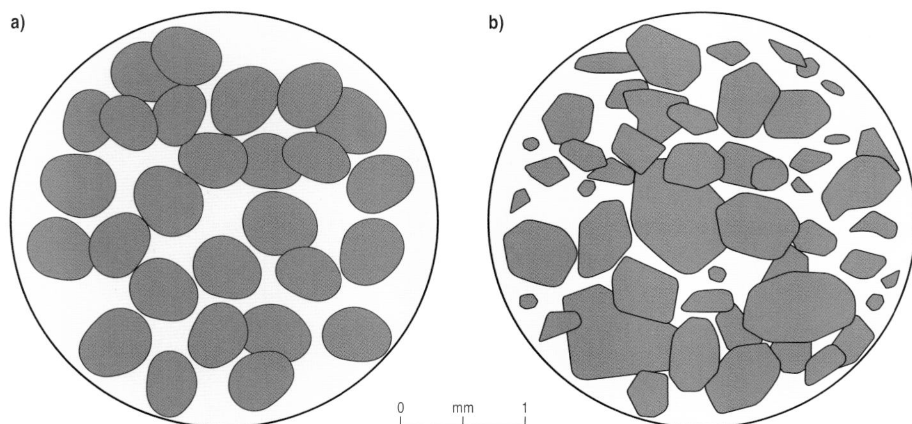

Figure 1.1 Sediment consisting of **A**) well sorted and rounded grains and, **B**) poorly sorted with subrounded to subangular grains. The degree of sorting may also indicate the energy, rate, and/or duration of deposition, as well as the transport process (river, debris flow, wind, glacier, etc.) responsible for laying down the sediment. Sorting of sediments can also be affected by reworking of the material after deposition.

the transportation of sediment, abrasion of grains can readily occur, where sharp corners and edges are worn smooth, a process known as **rounding** (Fig. 1.1). Rounding of sediment grains can indicate the distance and time involved in the transportation of the sediment from the source area to where it accumulates. The speed of rounding depends on hardness and composition of grains. For example, a soft mudstone pebble will obviously round much faster, and over a shorter distance of transportation, than a harder, more resistant sandstone pebble. The rate of rounding is also affected by the grain size and energy conditions.

The transportation of sediment grains, the action by gravity, and the effects of biological and chemical changes in the deposited sediment are all factors that control the architecture of the resulting sedimentary deposits. Any of these processes can lead to the formation of sedimentary structures, which are the visible signs of the various processes as preserved in the rock record. It is the job of the sedimentologist to interpret the sedimentary structures, and because particular processes are characteristic of different environments of deposition, the study of the sedimentary structures makes it possible to reconstruct environments of deposition and past palaeogeography.

Sedimentary environments

Understanding modern environments of deposition allows sedimentologists to understand the environments in which ancient sedimentary rocks were deposited and thereby help us recreate past conditions on the Earth. You may well have stood on a beach when the tide has gone out and admired the ripples that have been preserved in the sand. These ripples can then be used to identify that the sedimentary environment was a shallow marine beach setting, and this is especially important for ancient sedimentary settings (Fig. 1.2). Sediment that was eroded and transported will eventually be laid down layer by layer in a depositional

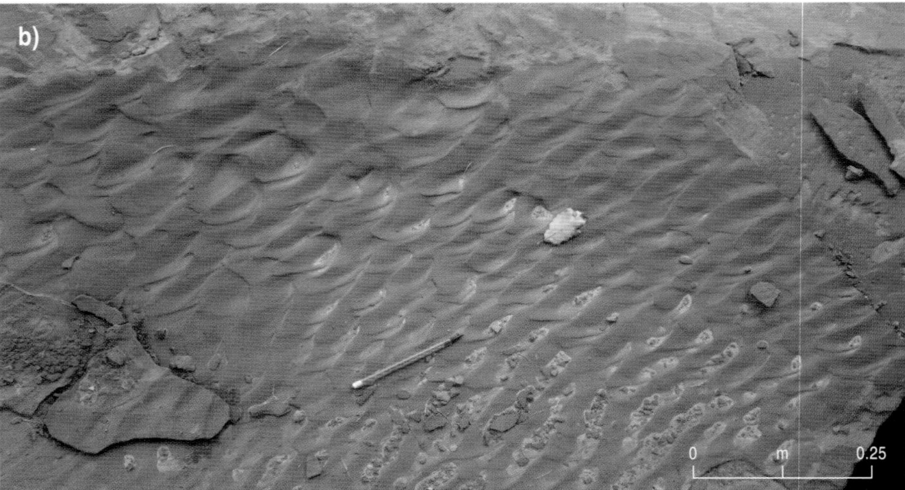

Figure 1.2 Modern and ancient ripples. **A**) Modern ripples, North Norfolk, UK; **B**) Ancient ripples from Triassic sandstones near Annan, Dumfries and Galloway, Scotland, UK.

environment, and over geological time will create the stratigraphic rock record.

Deposition of sediment may take place in many different environmental settings, and with most weathering and erosion taking place in continental mountainous regions, climate, topography, tectonic activity and local geology all control the amount, type and rate of sediment that is dispersed. Sedimentologists generally subdivide the depositional settings into three major depositional settings: continental (on the land), shoreline (beaches and deltas) and open marine, each with distinctive sediment characteristics that are used to compare with the geological (stratigraphic) rock record.

The Facies concept

The Swiss geologist and palaeontologist Amanz Gressly (1814–65) first introduced the term 'facies' in 1838 with his work on the Jura Mountains of central Europe. He discovered that individual layers or strata change their characteristics laterally: they are not uniform along their entire extent and undergo a facies change. A **facies** is a distinctive packet of sedimentary rock with specific sedimentary features that distinguish it from other facies (Fig. 1.3). Different facies are recognized based on sediment type (e.g. sandstone, limestone, coal), grain size, colour, texture, sedimentary structures and fossils. Sedimentary facies are the primary building blocks of all sedimentological studies. They provide the means of recording data, describing sediment and successions, and for making interpretations about the depositional processes.

Generally, facies are distinguished by sedimentologists based on what

Figure 1.3 Sedimentary facies of repeating sandstone and mudstone from Carboniferous deltaic sediments, Howick, Northumberland, UK.

aspect of the rock or sediment is to be studied. Thus, facies based on sediment characteristics such as grain size and mineralogy are called **lithofacies**. Lithofacies are used to interpret, correlate, and map sedimentary rocks. Facies based on fossil content are called **biofacies**. Particular facies may together form a characteristic **facies association** or may succeed one another to form a distinctive facies sequence.

Walther's Law of Facies, or simply Walther's Law, named after the geologist Johannes Walther, states that the vertical succession of facies reflects lateral changes in environment. Conversely, it states that when a depositional environment 'migrates' laterally, sediments of one depositional environment come to lie on top of another (Fig. 1.4). A classic example of this law is the vertical stratigraphic succession that typifies marine **transgressions** (sea-level rise) and **regressions** (sea-level fall).

Facies analysis is an important tool in the reconstruction of ancient environments of deposition and palaeogeography (*see* chapter 4). It has also proved invaluable in the prediction and exploitation of economic deposits.

Layer cakes and correlation

Stratigraphy is one of the fundamental disciplines of geology and is concerned with the study of rock layers (**strata**), which are usually of sedimentary origin. It seeks to interpret rock layers as sequences of events in the history of the Earth (often analogous to layers of a sponge cake; Fig. 1.5). Stratigraphy and sedimentology allow the sedimentologist to understand the Earth's surface in different places at different times and relate them to one another. It was **Nicolaus Steno** (1638–86; Fig. 1.6) who demonstrated the principle of superposition and showed that sedimentary rock strata were originally laid down horizontally on top of one another by the deposition of particles that fall out of suspension in a fluid (Fig. 1.7). Consequently, in any series of strata the youngest layers must be those at the top, and the oldest must lie on the bottom. He also determined that tilted and deformed strata result from displacement by the Earth's movements after deposition, and proved that fossils are the remains of once living organisms. These basic principles have ultimately become the 'stratigraphical column' in use by geologists today (Fig. 1.7). However, it was not until **William 'Strata' Smith** in the 1790s and early nineteenth century that there was the first practical large-scale application of stratigraphy. William Smith made the connection between fossils and the layer of rocks they were in, and used this to create the first geological map of England and Wales. He became known as the

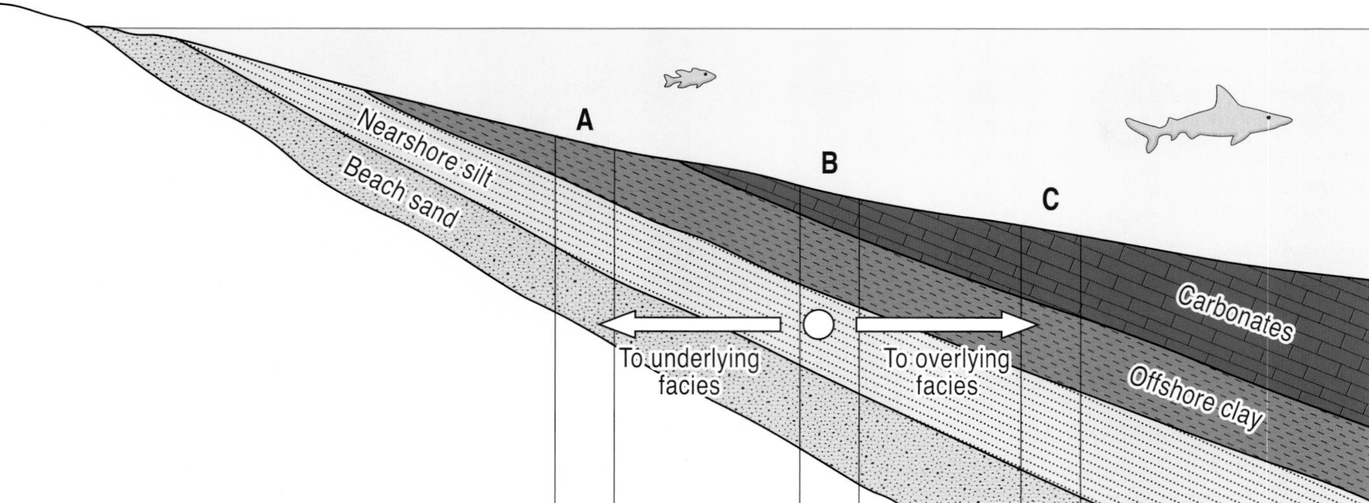

Figure 1.4 The principle of Walther's Law. The vertical sequence of sedimentary facies mirrors the original lateral distribution of sedimentary environments.

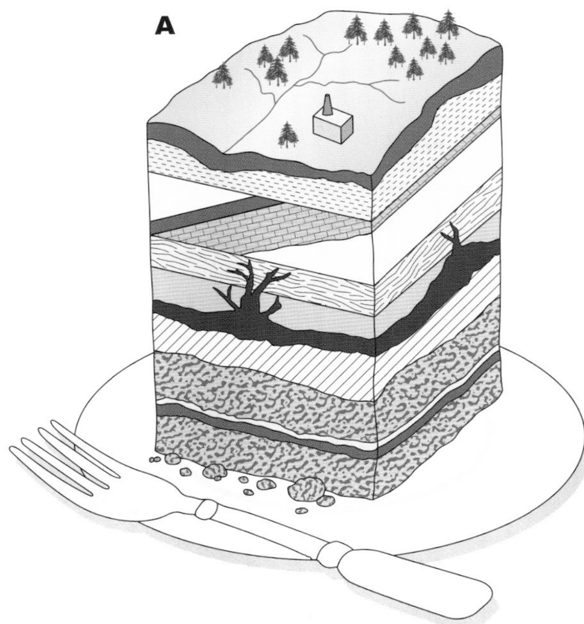

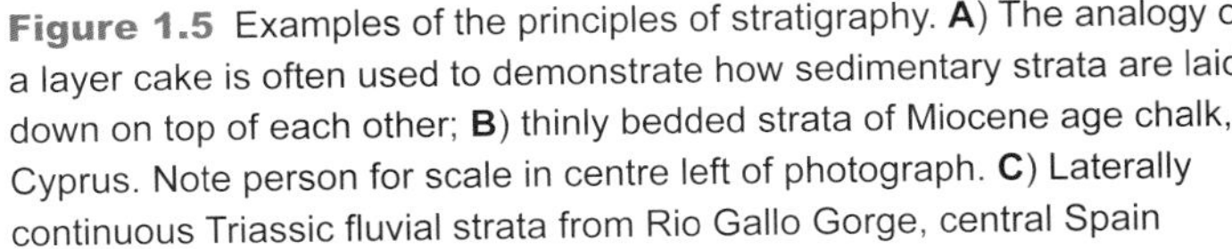

Figure 1.5 Examples of the principles of stratigraphy. **A**) The analogy of a layer cake is often used to demonstrate how sedimentary strata are laid down on top of each other; **B**) thinly bedded strata of Miocene age chalk, Cyprus. Note person for scale in centre left of photograph. **C**) Laterally continuous Triassic fluvial strata from Rio Gallo Gorge, central Spain

Figure 1.6 Portrait of Nicolaus Steno (1638–86). Nicolaus Steno was a Danish Catholic bishop and scientist who was a pioneer in understanding the basic principles of stratigraphy that underlie modern geology. Copyright http://nielssteensen.dk

'Father of English Geology'. Smith's map of 1815, called *A Delineation of the Strata of England and Wales with part of Scotland*, was the first geological map to identify the layers of rock based on the fossils they contained. Smith revolutionized the study of geological time and the order of the succession of life. Today, it is accepted that looking at fossils is the most accurate way of comparing sedimentary rocks and answering questions of geological time (*see* chapter 5).

Sedimentology and stratigraphy are frequently used together, as this provides an understanding of the sedimentary processes and resulting products, in space and through geological time. Past climate change is frequently recorded in the rock strata at local and global scales. Studies of past changes often reflect on the current situation

a)

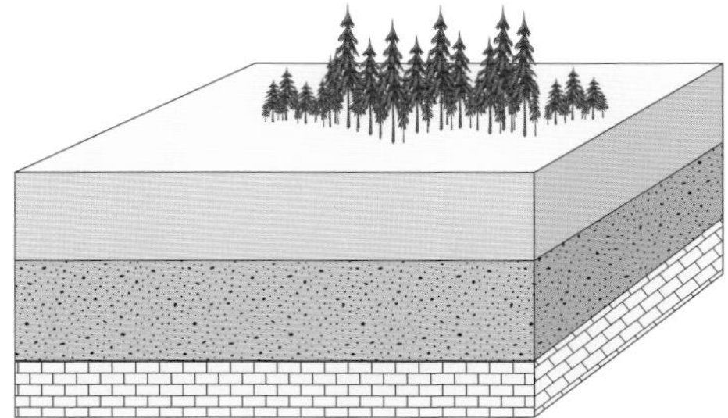

Principle of superposition

b)

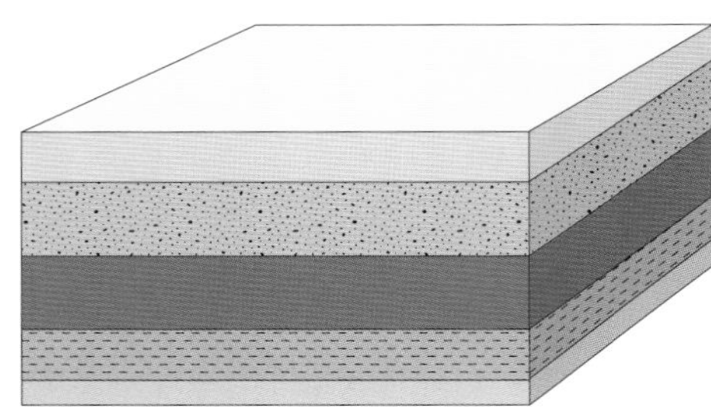

Principle of original horizontality

c)

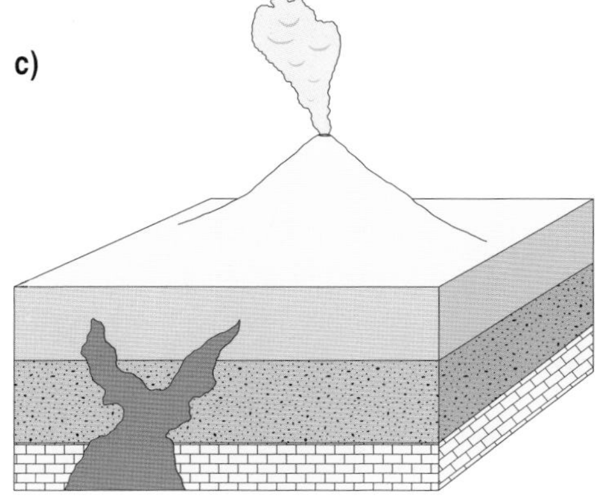

Principle of cross-cutting relationships

d)

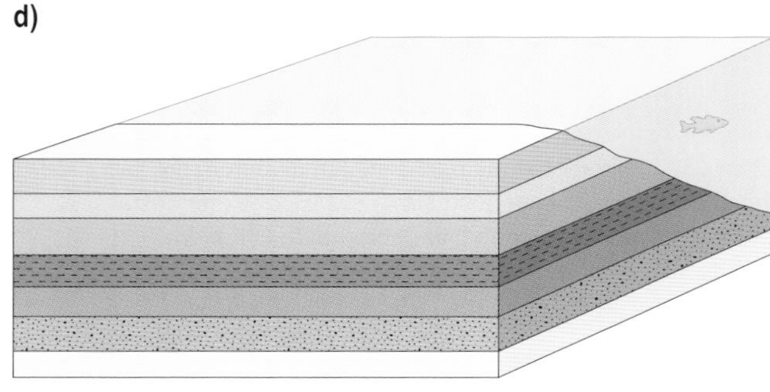

Principle of lateral continuity

Figure 1.7 Nicolaus Steno's four principles of stratigraphy. **A**) The **principle of superposition** states that when sediments are deposited, those which are deposited first will be at the bottom, and so the lower sediments will be the older; **B**) the principle of original horizontality states that sediment is originally laid down flat; **C**) an event that cuts across existing rock strata is younger than the original strata; **D**) strata can be assumed to have continued laterally far from where they presently end.

and provide a powerful tool in predicting future climate change. Perhaps the greatest advance in stratigraphy and sedimentary geology over the last 40 years has been with the development of **Sequence Stratigraphy**. Pioneered in the mid 1970s by Peter Vail and his co-workers at the US oil giant Exxon Mobil, it explains the complex geometries that sediments acquire as they accumulate in response to changes in rates of sedimentation, tectonics, and most importantly sea level. Sequence stratigraphy is a system for interpreting large-scale, three-dimensional arrangements of sedimentary strata and the factors that influence their geometrical relationships. It grew out of the subsurface analysis of continental margins for hydrocarbon exploration and is now common practice to apply sequence stratigraphic frameworks to both subsurface and field sedimentology.

The basic premise of sequence stratigraphy is that a change in sea level results in a change in patterns of sedimentation in most sedimentary settings and builds upon the basic long-established concepts of stratigraphy. The principal sedimentary unit in sequence stratigraphy is the **depositional unit** (or sequence) and is a package of sedimentary strata bounded above and below by major surfaces (**sedimentary sequence**). These are erosion surfaces (unconformities) and are technically defined as **sequence boundaries** (Fig. 1.8). Sequence boundaries are deemed the most significant surfaces in sequence stratigraphy and are formed as the result of a sea-level fall. The jargon associated with sequence stratigraphy can be quite baffling at first, and the detail is beyond the scope of this book. However, once important sequences and sequence boundaries have been recognized, then these provide a 'key' for correlating and integrating seismic, well and outcrop data, allowing the chronological order of **sedimentary basin** filling and erosional events to be determined – which helps to make stratigraphy a piece of cake!

The hole in the ground to place the dirt

Sedimentary basins are bowl-shaped depressions or topographic lows of the Earth's crust where sediment can accumulate into successions hundreds to thousands of metres thick. They are dustbins of weathering, erosion and transportation of sediments, and their sedimentary fill provides unique evidence for the environmental conditions that occurred during the basin's lifetime. Each basin has a stratigraphy partly controlled by local conditions and also by global scale processes such as tectonic activity, climate changes and sea-level fluctuations. Sedimentology has a vital role, for it is largely the evidence provided in the sedimentary rocks that provides data for reconstructions of earth evolution. Most sediment studied by sedimentologists is found within sedimentary basins, which is where sedimentary environments are best reconstructed.

The main control on producing sedimentary basins is tectonics, because tectonic events produce areas of subsidence that become filled with sediments. **Subsidence** is the local and regional scale change in the Earth's crust in the form of a downward shift

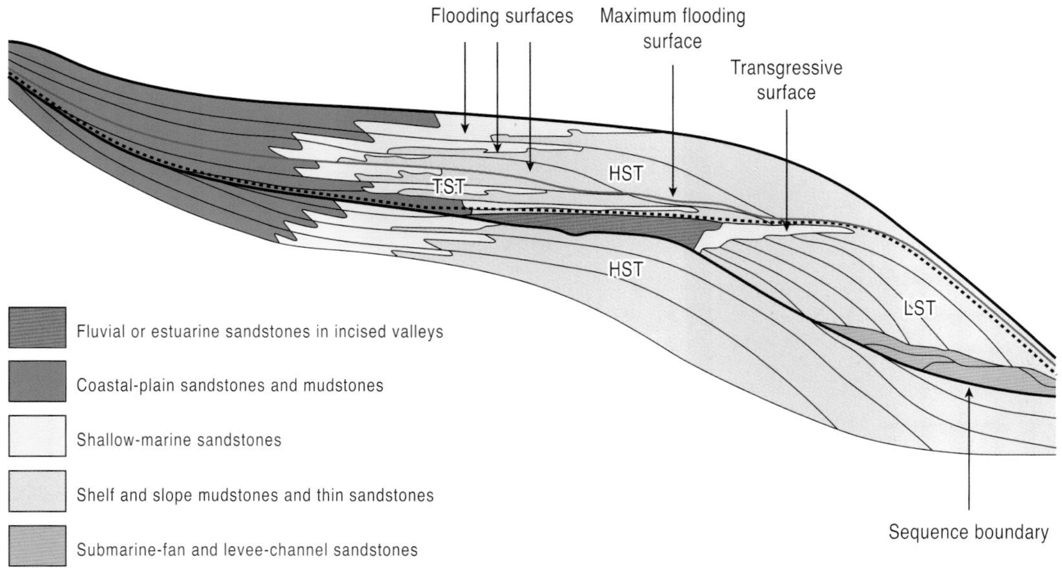

Figure 1.8 An idealized depositional sequence. Sediment is supplied from the left of the figure. The sediment thicknesses are greatly exaggerated. Boundaries between different stratal packages represent former positions of the land surface and sea floor. The sedimentary stratal packages of different sedimentary environments migrate landward and seaward according to sea-level rise and fall. Figure adapted from: Van Wagoner, J.C., Mitchum, R.M., Campion, K.M., Rahmanian, V.D. (1990) Siliciclastic sequence stratigraphy in Well Logs, Cores, and Outcrops: Concepts for High-Resolution Correlation of Time And Facies. *AAPG Methods in Exploration Series no. 7.*

relative to a datum, usually sea level. Continued sedimentation while the basin is tectonically active and subsiding will often affect the distribution and style of sedimentation. It is in sedimentary basins that sequence stratigraphy can be best applied, as any particular basin, whether active or inactive, will possess its own unique sedimentary architecture.

Different kinds of sedimentary basins are defined on the basis of the movement of the tectonic plates and can be divided into five simple groups of basin settings (Fig 1.9):

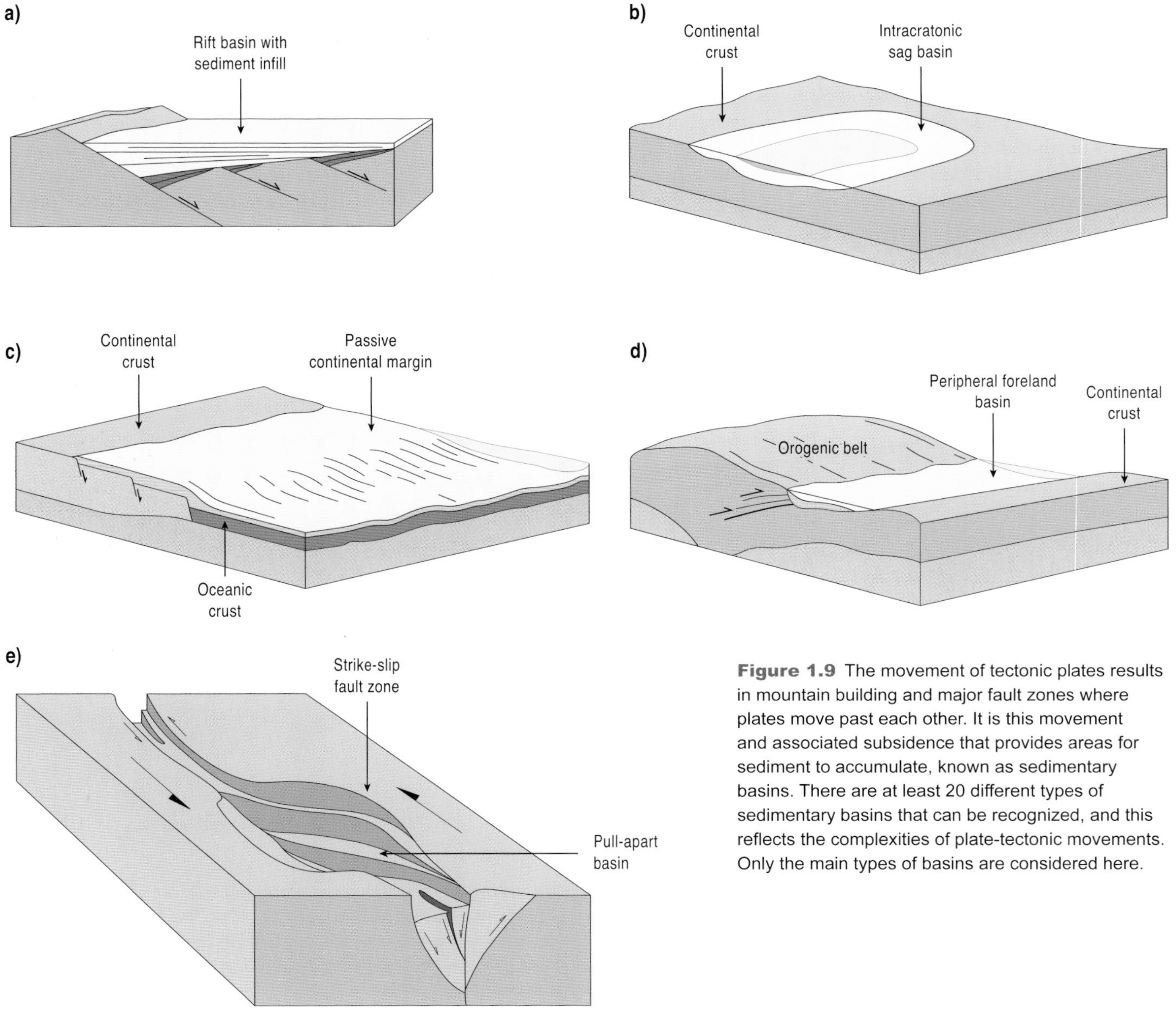

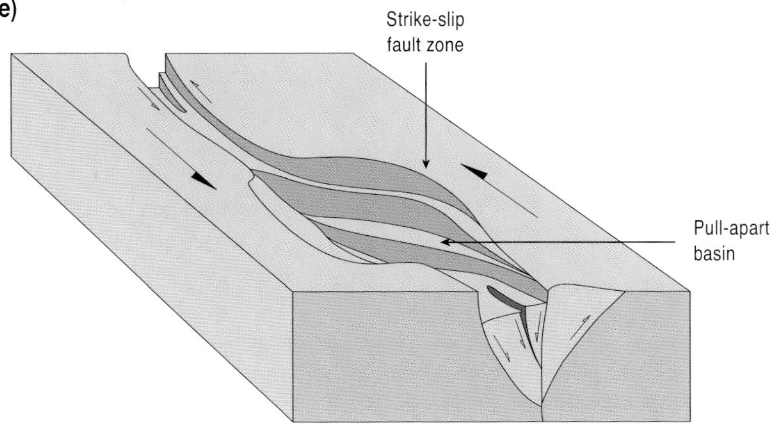

Figure 1.9 The movement of tectonic plates results in mountain building and major fault zones where plates move past each other. It is this movement and associated subsidence that provides areas for sediment to accumulate, known as sedimentary basins. There are at least 20 different types of sedimentary basins that can be recognized, and this reflects the complexities of plate-tectonic movements. Only the main types of basins are considered here.

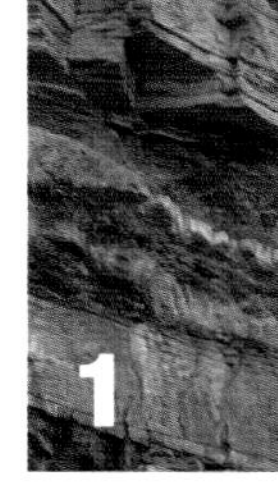

1 Rift basins

These basins form where the continental crust is being stretched. During the early phases of rifting the Earth's surface subsides because it thins as it is stretched. A similar analogy can be achieved by using a chocolate bar with a caramel centre and gently stretching from either end. As the rift grows, slip on the border faults drops blocks of crust down, producing low areas and narrow mountain ridges. Rifts can occur in both continental and marine settings. The East African Rift System is a good example of modern day rift with many rift basins located along its length. The rift is a narrow zone in which the African Plate is in the process of splitting into two new tectonic plates, and will eventually produce a new ocean. Rift basin sediments can be continental or marine in origin.

2 Intracratonic basins

These tend to develop in the interiors of continents away from tectonic plate margins and mountain belts. They have generally round or oval shapes, and have long geological histories of relatively slow subsidence. One hypothesis on their formation is that they are thought to form over areas of previous rifting. When rifting stops, the once hot and stretched crust starts to cool, contracts and sinks. This sinking is known as thermal sag and results in an intracratonic basin. They tend to be filled with continental sediments, although flooding from an adjacent ocean can result in large **epicontinental seas**.

3 Passive margin basins

These form along the margins of continents that are not tectonic plate boundaries. They are usually underlain by a former rift with oceanic crust. Passive margin basins occur because long after the rifting has ceased, the thermal relaxation and subsidence continues. The sediments associated with passive margins can be both **carbonate** and **clastic** and develop thicknesses of 10–20 km. It was in passive margin basins that sequence stratigraphy was developed and is still most successfully used. A good example of a modern passive margin basin is the Gulf of Mexico margin along the southern United States.

4 Foreland basins

Foreland basins are associated with regions of compressional tectonics and form adjacent and parallel to mountain belts. They are formed primarily as a result of the downward flexing of the ***lithosphere*** in response to the weight of the adjacent mountain belt, though many geological and geodynamic processes combine to control their subsequent evolution. The foreland basin receives sediment that is eroded off the adjacent mountain belt, filling with thick sedimentary successions that gradually decrease in thickness away from the mountain belt. In general, sediments deposited in foreland basins tend to show a transition from deep marine to continental environments and can reach thicknesses in excess of 10 km. The Persian Gulf is a good example of a foreland basin produced by the Zagros Mountains of Iran (Fig. 1.10).

5 Strike-slip basins

Strike-slip faults are classified by a horizontal sense of movement along the fault plane. In such cases there will not be any uplift or basin formation. However, strike-slip fault planes are seldom straight, and areas of localized compression alternate with extension and can give rise to varying sized basins. Strike-slip basins tend to be filled with either continental or marine sediments according to the climate and local geological setting. Examples of modern strike-slip basins occur along the San Andreas Fault of California, USA and the North Anatolian Fault of Turkey.

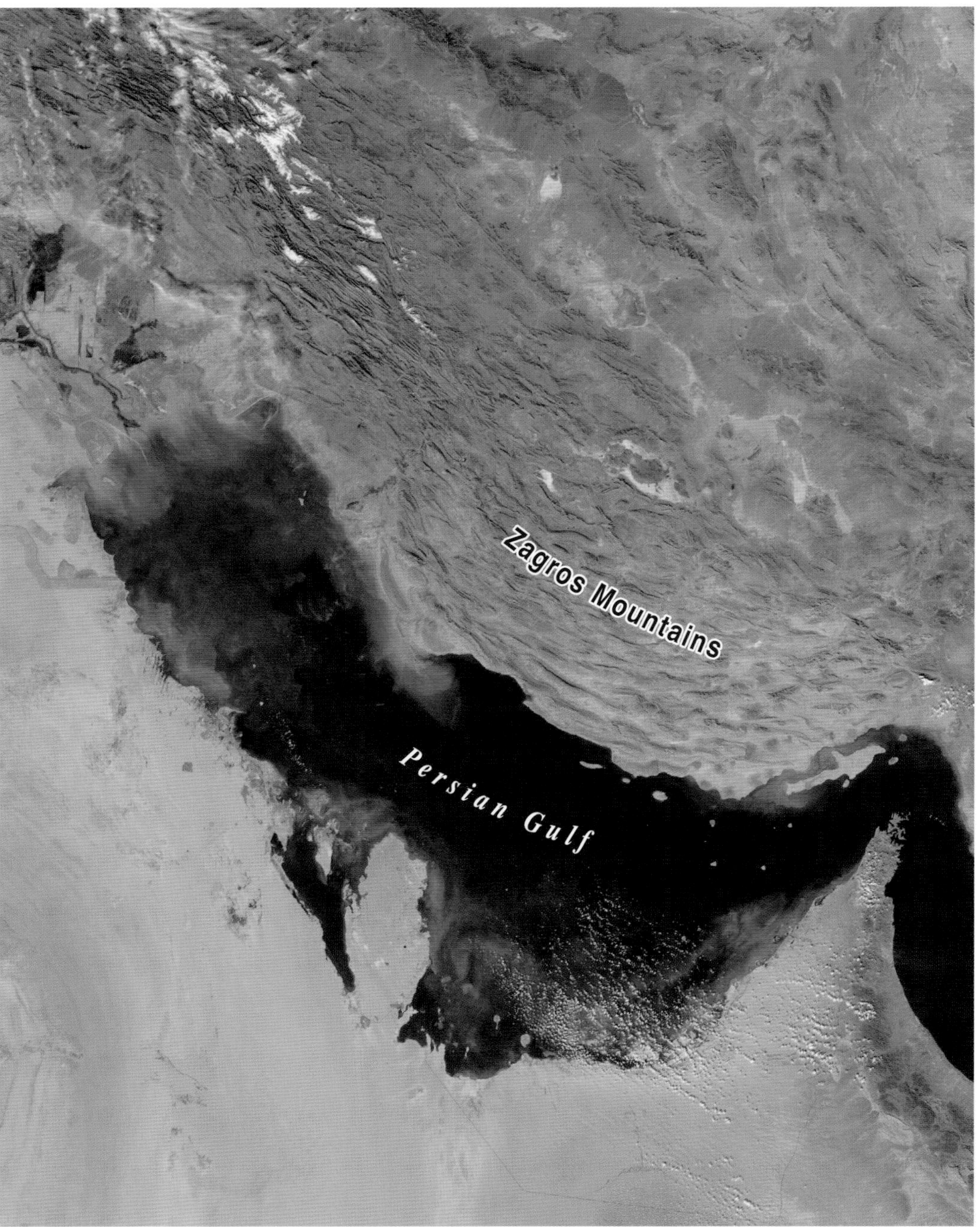

Figure 1.10 The Persian Gulf and the area of lowland occupied by the alluvial plain of the Tigris–Euphrates river system represent the active foreland basin to the Zagros Mountains, caused by the loading of the leading edge of the Arabian Plate by the Zagros thrust sheets. Captured by the Moderate Resolution Imaging Spectroradiometer (MODIS) on NASA's Aqua satellite on November 28, 2007. Copyright NASA.

2 Sediment to sedimentary rock

There has been sediment on Earth since the Archaean, with the oldest known sediment grains dating from at least 4400 million years ago. Sediments are interwoven into the 'fabric' of the Earth with at least 70% of the rocks on the Earth's surface being sedimentary in origin. The atmosphere, hydrosphere and biosphere all play their roles in the creation of sediment, and in it becoming a sedimentary rock. The frequently used conceptual model of the **rock cycle** is a useful visualization to see the changing relationships between the three major rock types of igneous, metamorphic and sedimentary rocks (Fig. 2.1). Sediment and sedimentary rocks can yield important information about igneous and

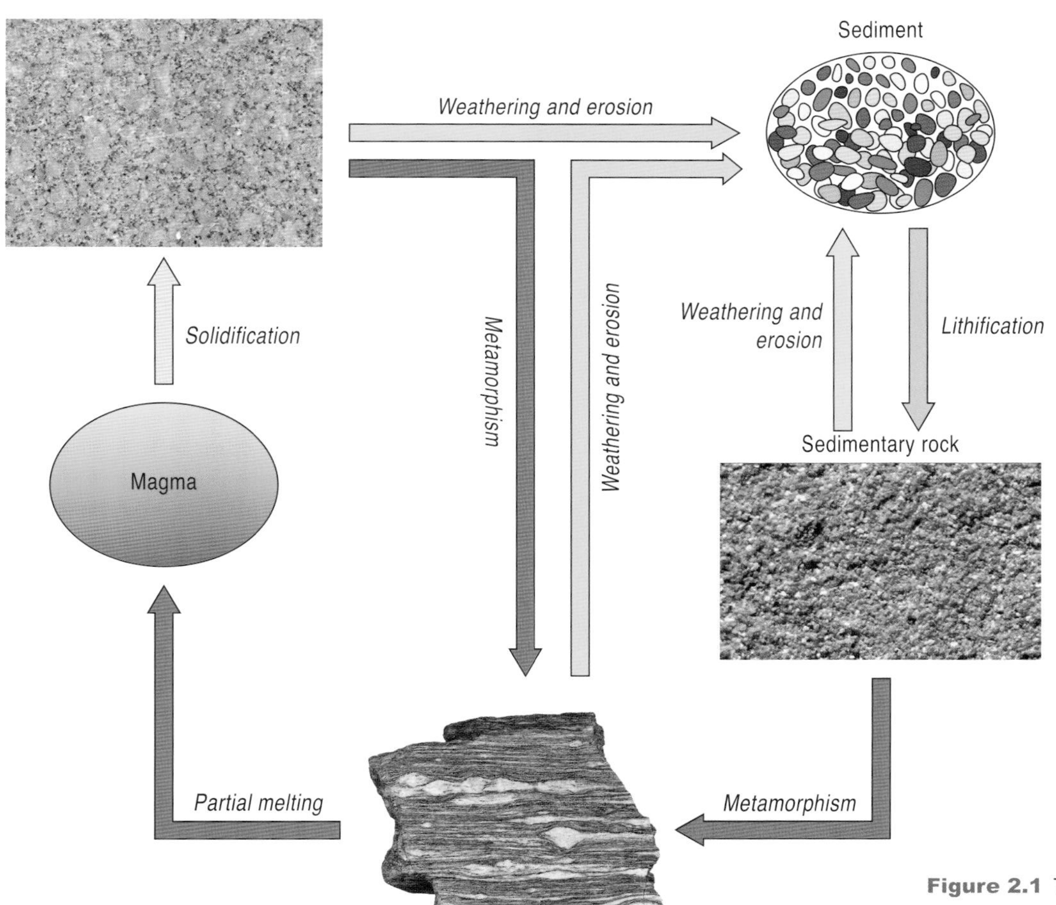

Figure 2.1 The rock cycle illustrating the role played by sediments and sedimentary rocks.

metamorphic rocks that have long since been eroded. Nevertheless, sedimentary rocks are uniquely important because they contain a geological record and the bulk of the Earth's energy resources.

Sediment

A sedimentologist refers to the kind of deposits described as sand, mud, gravel, and even shell fragments as sediment. Sediment is a collective term for loose fragments of rocks or minerals that originate from weathering and erosion of pre-existing rocks or from direct precipitation out of solution, e.g. directly from seawater. Sediments tend to form a thin veneer over much of the Earth's surface, and when sediments are initially deposited they are considered to be unconsolidated, where each grain is separate and they are not cemented to each other.

Most sediment is transported some distance by water, ice, wind or gravity. During transportation the sediment continues to be weathered and to change character. Most commonly as they move, rock fragments and grains hit and scrape each other, and over longer distances can remove corners and angular edges. This is known as **rounding** and usually sediment grains close to their source are less well rounded than those that have been transported some considerable distance (Fig. 2.2). Associated with rounding of grains is **sorting**. Sorting is the process by which sediment grains are selected and separated according to size (or grain shape or specific gravity) during transportation. Glaciers transport sediment of all sizes and glacial sediments are considered to be *poorly sorted*. In comparison, wind-blown sediments of a sand dune are *well sorted* as the grains are nearly all the same size (Fig. 2.2).

Eventually sediment will come to rest and deposition occurs. The sediment is deposited when the transportation process can no longer carry its load. However, not all sediments are preserved. Quite often sediment is only temporarily deposited and then is reworked by the same transportation processes. This is very common in rivers, where sediment can be reworked several times due to flood events. In general, most sediments are likely to be preserved if they are deposited in a subsiding sedimentary basin (see chapter 1) and if they are covered or buried by later sediments.

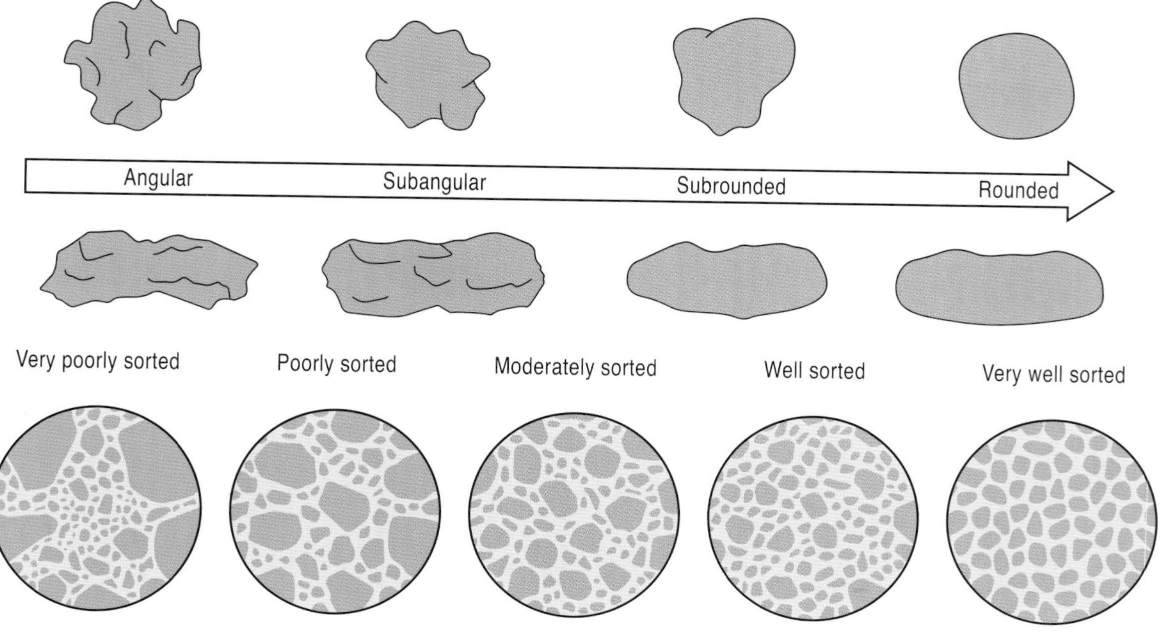

Figure 2.2 Rounding and sorting of grains allows the sedimentologist to infer distances grains have travelled from source area and even allow likely modes of travel to be interpreted.

Sediment to rock

Once sediment has been deposited over time it will undergo the process of **lithification**, where loose, unconsolidated sediment changes into a sedimentary rock. Most sedimentary rocks become lithified through a combination of **compaction**, where loose grains are packed together more tightly, and **cementation,** in which the precipitation of a new mineral around the sediment grains binds them into a firm, coherent rock (Fig. 2.3). Common cementing minerals are calcite ($CaCO_3$), **quartz** (SiO_2), iron oxides and clay minerals. Some sedimentary rocks are created from direct precipitation out of solution, without passing through the loose-sediment stage, but still can undergo compaction and further cementation.

Sedimentary rocks

Sedimentologists subdivide sedimentary rocks into five major classes, based on their mode of origin, although no classification is perfect. 1) **Clastic sedimentary rocks** (also termed terrigenous or siliciclastics) consist of cemented together fragments and grains derived from pre-existing rocks. 2) **Biogenic sedimentary rocks** are derived from the skeletal remains and soft organic matter of pre-existing organisms. 3) **Organic sedimentary rocks** are composed of organic carbon compounded from relicts of plant material. 4) **Chemogenic sedimentary rocks** (also termed chemical or authigenic) are those formed from the direct precipitation of minerals from a saturated solution. 5) **Volcanoclastic sedimentary rocks** are composed of grains and fragments derived from volcanic activity.

It is hard to estimate the relative proportions of the different sedimentary rocks on Earth, but it is roughly thought that between 70 and 85% of all sedimentary rocks are clastics (e.g. sandstones), whereas 15–25% are made up of biogenic sedimentary rocks (e.g. limestones or carbonate rocks). The other kinds of sedimentary rocks, although important for differing reasons, occur only in minor quantities.

Clastic sedimentary rocks

Clastic sedimentary rocks are one of the most diverse groups, ranging from coarse-grained conglomerates and breccia through sandstones to fine-grained mudrocks. Clastic sedimentary rocks are classified based upon the size of individual rock fragments and/or grain size (Table 2.1).

Conglomerates and **breccias** are coarse-grained (>2 mm) sedimentary rocks formed through the cementation of rounded or angular gravel respectively. Because grains are rounded fairly rapidly during transportation, the angular fragments of a breccia are unlikely to have moved very far from their source. In comparison, the more rounded fragments of a conglomerate will have undergone a greater amount of transportation and be further from source.

Perhaps one of the mostly commonly recognized of all sedimentary rocks is **sandstone**. Sandstones are characterized as medium-grained sedimentary rocks that contain more than 50% sand size (0.063 mm to 2 mm) grains. The most common mineral in sandstone is quartz, the most commonly occurring mineral on the Earth's continents. The average composition of sandstone is about 65% quartz, but frequently

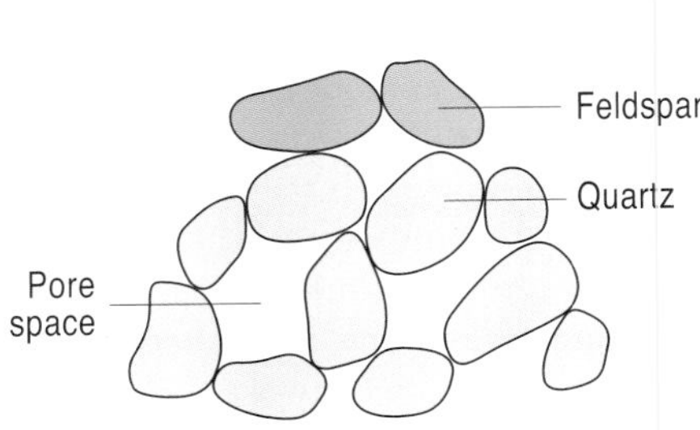

a) At time of depostion

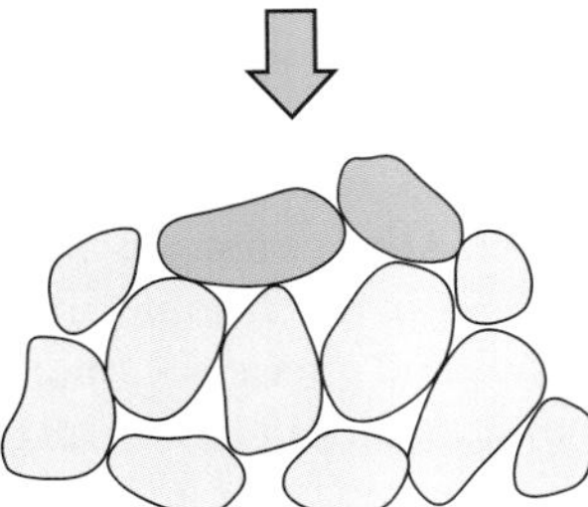

b) Compaction due to burial

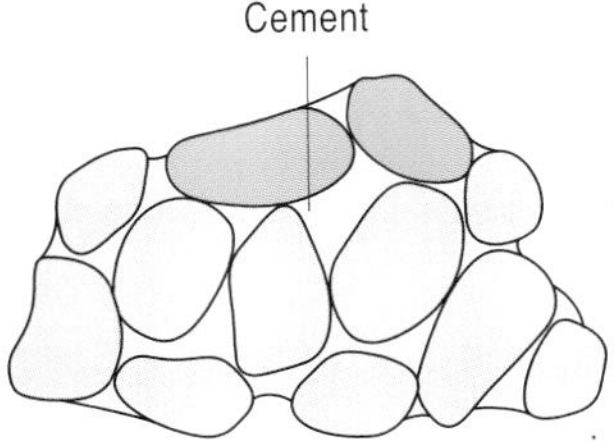

c) Cementation

Figure 2.3 The process where loose sand grains made from the mineral quartz (**A**) become lithified through compaction and cementation (**B**,**C**) to form a sandstone rock.

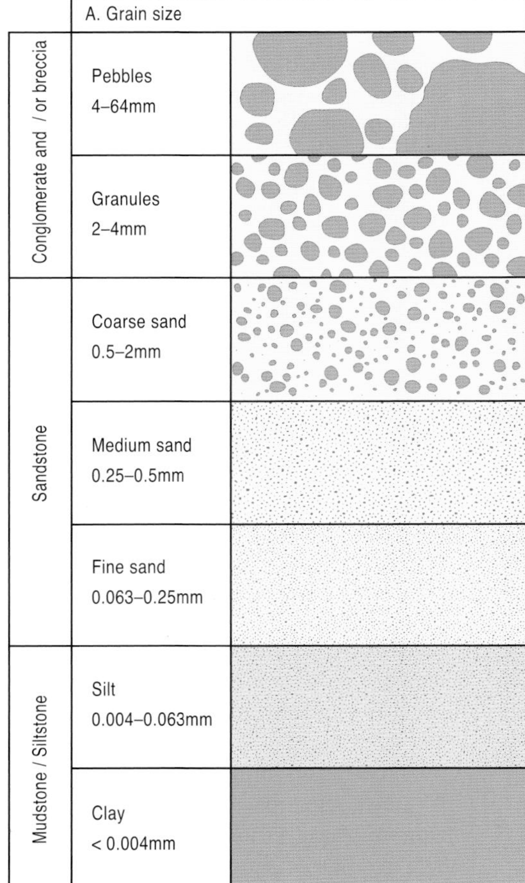

	A. Grain size	
Conglomerate and / or breccia	Pebbles 4–64mm	
	Granules 2–4mm	
Sandstone	Coarse sand 0.5–2mm	
	Medium sand 0.25–0.5mm	
	Fine sand 0.063–0.25mm	
Mudstone / Siltstone	Silt 0.004–0.063mm	
	Clay < 0.004mm	

Table 2.1 Classification of sediments and sedimentary rocks by grain size.

a) Quartz arentie: pure quartz

b) Arkose: feldspar–rich

c) Lithic sandstone: rock-fragment–rich

d) Greywacke: matrix-rich

0 mm 2

Feldspar · Rock fragments · Quartz grains

Figure 2.4 Classification of sandstones based on different types of grains and the type of matrix present between the grains.

it is greater than 95%. Many different types of sandstone occur, and they are classified based on their mineral compositions. There are four commonly recognized sandstone types (Fig. 2.4). The most compositionally mature are **quartz arenites** with at least 95% quartz. In addition, their grains are usually well sorted and rounded, demonstrating that these sandstones are the products of extensive sediment reworking, so that all grains other than quartz have been broken down and removed from the sandstone. Sandstones that contain at least 25% **feldspar** are known as **arkoses.** Arkoses are sandstones derived from feldspar-rich source rocks, usually granites and gneisses. **Litharenites** account for about 25% of all the different sandstone types and usually have rock fragment content in excess of feldspar. Litharenites are usually considered to be poorly sorted, and grains may be subrounded to subangular. The immature sandstone composition implies short to moderate transport distances and quite often high rates of sedimentation, such as in deltaic depositional environments. Finally, **greywackes** are sandstones with at least 15% fine-grained muddy matrix between the quartz grains. Most greywackes probably formed from sediment gravity flows called **turbidity currents** (Fig. 2.4d).

Mudrocks make up at least 50% of all sedimentary rock successions and are defined as fine-grained clastic rocks, composed of grains <0.0063 mm (63 μm). Common synonyms used are shales, mudstones and claystone. Due in part to their grain size and vulnerability to weathering, this class of rocks has received much less focus of attention by sedimentologists than other types. However, this is changing, partly with the growth of shale gas exploration (see chapter 6).

Biogenic sedimentary rocks

Off the coast of Queensland, Australia lies the world's biggest single structure made by living organisms, known as the Great Barrier Reef. It contains an incredibly diverse community of corals and algae, around which live other creatures such as oysters, clams, and snails (molluscs), and plankton float in the seawater above. All of these creatures may not seem to have much in common, but many of them have shells made from the mineral calcite (or its polymorph aragonite, $CaCO_3$). When these organisms die some may stay *in situ,* (e.g. coral reef builders) or some may be moved and broken by currents, as commonly occurs with molluscs. A sedimentary rock formed by skeletal fragments of calcite or aragonite is known as **limestone** (Fig. 2.5). However, sedimentologists use the term limestone for a wide range of 'carbonate-rich' sedimentary rocks that form in a variety of settings, regardless of origin. For example, limestones can also form from direct organic precipitation from seawater or even from geothermal hot springs, forming a crystalline limestone known as travertine (Fig. 2.6).

The direct precipitation of carbonate from warm, supersaturated, shallow, highly agitated marine waters leads to spherical to sub-spherical sand-size grains (typically 0.2–0.5 mm in diameter) called **ooids**. The mechanism of formation starts with a small fragment of sediment acting as a 'seed', e.g. a piece of a shell. Strong currents wash the 'seeds' around on the seabed, where they accumulate layers of chemically precipitated calcite from the supersaturated water.

Figure 2.5 Crinoidal limestone. Crinoids are sea animals that had long stems, cup-like bodies and long filter arms, and are related to sea urchins and starfish. They were particularly important and diverse throughout the Palaeozoic (Cambrian to Permian) and their remains have contributed substantially to creating thick sequences of crinoidal limestones.

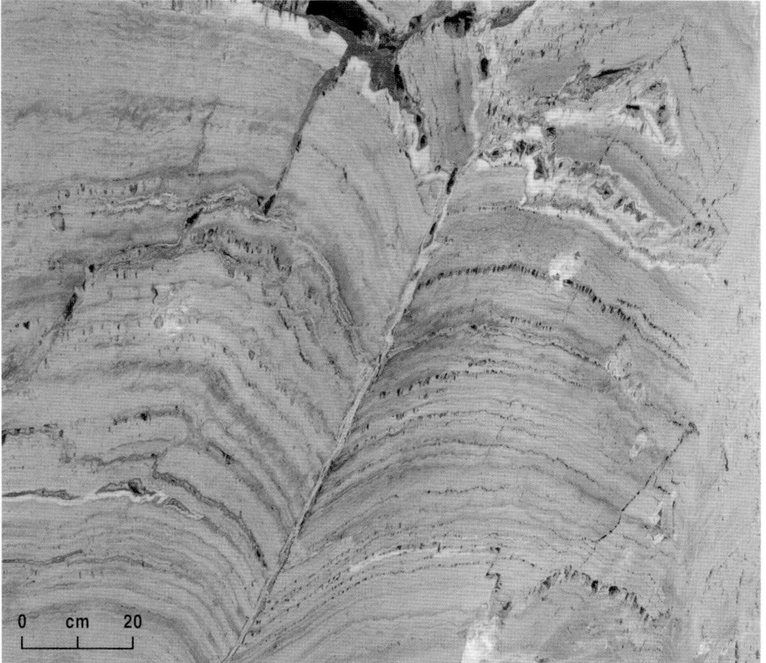

Figure 2.6 Travertine, Central Iran.

Carbonate sediment composed of ooids is referred to as **oolite** (Fig. 2.7). An ancient example of an oolite is the Middle Jurassic age (135 million years old) decorative Bath stone that is used for many of the buildings in the World Heritage City of Bath, England (Fig. 2.7).

Dolomite differs from limestone in that it contains the mineral dolomite ($CaMg[CO_3]_2$). Most dolomites are formed from the replacement of limestone, forming through the chemical reaction of calcite (limestone) with magnesium-bearing groundwater. This process of dolomitization often occurs soon after burial and can result in the complete obliteration of any primary sedimentary structures or fossils that the limestone may have contained at the time of deposition. Dolomites can be found throughout the geological rock record, with none more spectacular that the Dolomite Mountains of Northern Italy (Fig. 2.8). The Dolomite Mountains have inspired voluminous research into the origin of dolomite, questioning whether it is a primary precipitate or a secondary replacement product. Recently, with the recognition that microbes can

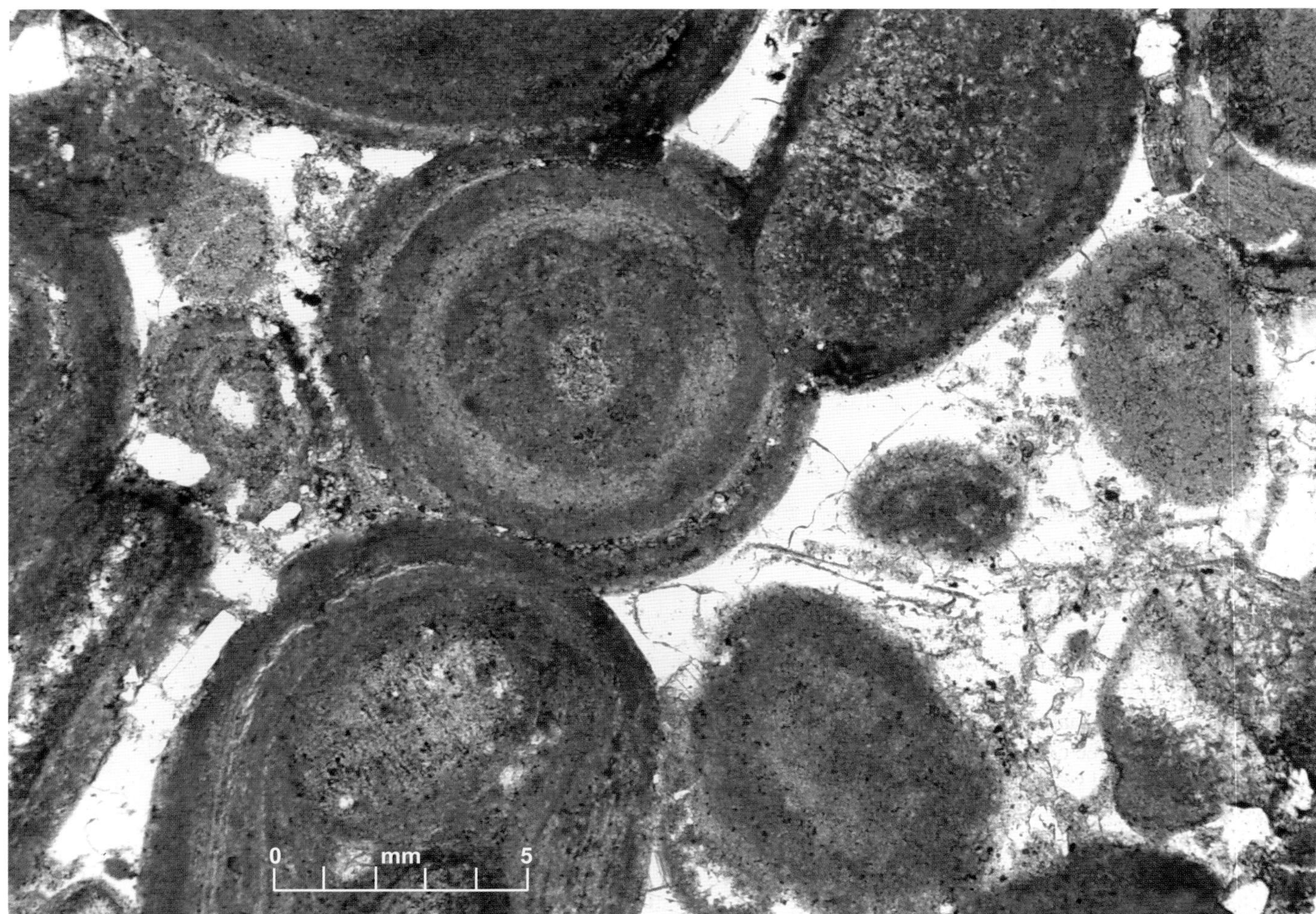

Figure 2.7 Photomicrograph of oolitic limestone made up largely of sand-sized, rounded pellets of calcium carbonate, which are formed in warm, shallow water where carbonate sediment is moved about by currents. Jurassic, North Yorkshire, UK.

Figure 2.8 'The Dolomites' are a mountain range in north-eastern Italy, and part of the Italian Alps. It is one of the largest exposures of dolomite rock on Earth – from which the name is obtained. The Langkofel, South Tyrol, Italy.

mediate dolomite precipitation, a new geomicrobiological approach is being used to explain the origin of dolomite.

Organic sedimentary rocks

Coal is an organic sedimentary rock that results from the compaction and burial of plant material that has not completely decayed. Coals are brown to black, soft to hard, low density, organic carbon-rich sedimentary rocks that have long been mined due to the economic importance of coal for energy generation (Fig. 2.9; see chapter 6).

Not all organic rocks are derived from plant material. Organic material can form from decomposition and decay of planktonic organisms, algae, spores and pollen. The organic material can mix with muds and be incorporated into mudrocks known as

Figure 2.9 Anthracite coal is the hardest of all coal types and has the highest carbon content (92–98% carbon).

oil shales or black shales. The continued burial of such organic-rich rocks may result in oil and gas, the details of which are discussed in chapter 6.

Chemogenic sedimentary rocks

Evaporites are perhaps the most widely recognizable of all the chemogenic sedimentary rocks. Evaporites form directly from the precipitation of crystals from water following the concentration of dissolved salts by evaporation. The main evaporite minerals that are commonly encountered include gypsum, anhydrite and halite. All evaporates have a crystalline texture, and crystal size can vary considerably depending on the concentration of the dissolved salts in the sea, lagoon or lake depositional settings. Evaporites are particularly prone to diagenetic modification following burial, with dissolution, recrystallization and deformation common. There are relatively few modern depositional environments of evaporite sediments (Fig. 2.10), but in the geological past huge volumes of evaporites have been precipitated, reaching more than 1.5 km in thickness, for example during the Permian in the Zechstein Basin of NW Europe.

Volcaniclastic sedimentary rocks

In many volcanically active regions of the World, sediments can be composed mainly of grains and clasts derived from local volcanic activity. Volcaniclastic sediments are best classified in terms of their modes of formation, as described in Table 2.2. The most widely known of all volcaniclastic sediments are the pyroclastic flows and fall deposits (Table 2.2). Pyroclastic flows

Figure 2.10 The Dead Sea. **A**) Landsat image of the Dead Sea. **B**) Salts precipitated from the hypersaline waters. **C**) Layers of salts precipitated over geological time.

Table 2.2 Classification of the principal types of volcaniclastic sediments. These are composed principally of sediments of volcanic origin and usually derived from a contemporaneous volcanic event. Thanks are due to Richard Brown for advice on volcaniclastic sediments.

VOLCANICLASTIC SEDIMENTS		
Sediment types	**Subdivision**	**Origin and process**
Pyroclastic fall deposits	>64 mm Agglomerate (blocks, bombs and volcanic breccia) <64 mm Lapilli tephra (Lapillistone) <2 mm Coarse ash (pyroclastic sandstone) <0.063 mm fine ash (pyroclastic mudstone)	Result from the fall-out of volcanic fragments such as ash, tuff and tephra ejected from the vent of a volcano. They can be deposited in water and/or fall through air.
Pyroclastic flow deposits	Surge and flow deposits (ignimbrites)	Fast-moving currents of hot gas and rock fragments (collectively known as tephra), that move rapidly down the slopes of a volcano in response to gravity.
Epiclastic sediments	Epiclastic conglomerate, sandstone and mudstone Lahar deposits	Volcanic sediments that are produced by erosion of volcanic rocks by the action of wind, water, and / or ice, and by debris flows
Hydroclastic deposits	Hyalotuffs (explosive) Hyaloclastites (non-explosive)	Formed by lava fragmentation through contact with water; subaqueous eruptions and lavas flowing into water (e.g. the sea or a lake).
Autoclastic deposits	Clast-supported or matrix-supported volcanic breccias.	Often poorly sorted with angular breccias formed by autobrecciation of lavas as they flow down a volcano

normally hug the ground and travel downhill, or spread laterally under gravity away from a volcanic eruption. Their speed depends upon the density of the current amount of gas and rock debris, the volcanic output rate, and the gradient of the slope. Pyroclastic flows that contain a much higher proportion of gas to rock are known as pyroclastic surges. The lower density sometimes allows them to flow over higher topographic features such as ridges and hills. In comparison **pyroclastic fall** deposits tend to be better sorted and the deposits can drape over pre-existing topography and maintain a uniform thickness over relatively short distances (Table 2.2). Pyroclastic fall deposits vary in size and are dependent upon the volcanic eruptive style, with volcanic blocks and bombs associated with the most explosive of eruptions. Ash and tephra are frequently found interbedded amongst sediment successions and provide a useful correlative marker, and also radiometric ages, especially in non-marine sediments where there are frequently few fossils preserved and little to correlate or date (see chapter 1).

A **lahar** is a type of volcanic mudflow or debris flow composed of a slurry of pyroclastic material, rocky debris and water (Table 2.2). The mixture flows down from a volcano, typically along a river valley, and the amounts of water and rock debris it carries constantly change. Eruptions may trigger one or more lahars directly by quickly melting snow and ice on a volcano or ejecting water from a crater lake. More often, lahars are formed by intense rainfall during or after an eruption, and rainwater can easily erode loose volcanic rock and soil on hillsides and in river valleys. This is particularly the case for Mount Pinatubo, in the Philippines. Within hours of Mount Pinatubo's explosive eruption on 15 June 1991, heavy rains began to wash the ash and debris deposited by this explosion down into the surrounding lowlands in the form of giant, fast-moving lahars. By 1993, lahars had already caused more devastation in the lowlands than the eruption itself, and they continue to be a major hazard in the region.

Volcanism is an important contributor of mud- and sand-sized material in many different sedimentary environments, and its contribution to the sediment in deep marine basins is perhaps often overlooked. The deep-sea sedimentary environment is very calm, far removed from the storms that can disturb shallow shelf waters or rivers that drain the continents. The low energy of the environment is ideal for volcanic particles, which, typically consisting of fine ash, can be a rich source of sediment, particularly in volcanic regions.

Sediments in a pressure cooker!

The study of sediments and sedimentary rocks does not stop at those that have remained at the surface or in a sedimentary environment. In fact, considerable valuable information can be yielded from the sedimentary rock that has been buried often to a depth of several kilometres. The term **diagenesis** is used for all the physical, chemical and biological processes that occur during burial but prior to the onset of metamorphism. The term diagenesis is normally used following the initial process of lithification where sediment hardens into a rock. Diagenesis involves physical compaction of components due to pressure increase on burial, the precipitation of mineral cements from pore fluids, and phase transformations of mineral components.

The progressive burial of sediment leads to dewatering and compaction with grains becoming more closely packed. In mudrocks there can be up to a 90% reduction in thickness during burial, but this tends to be much less, between 40 and 50%, in coarser-grained sediments (Fig. 2.11). If compaction continues with even greater thicknesses of overlying sediment and/or tectonic pressure, then pressure dissolution can result, where the faces of grains dissolve where they are squeezed against neighbouring grains (Fig. 2.12). A sedimentary rock that continues to be deeply buried will experience ongoing diagenesis, both as the result

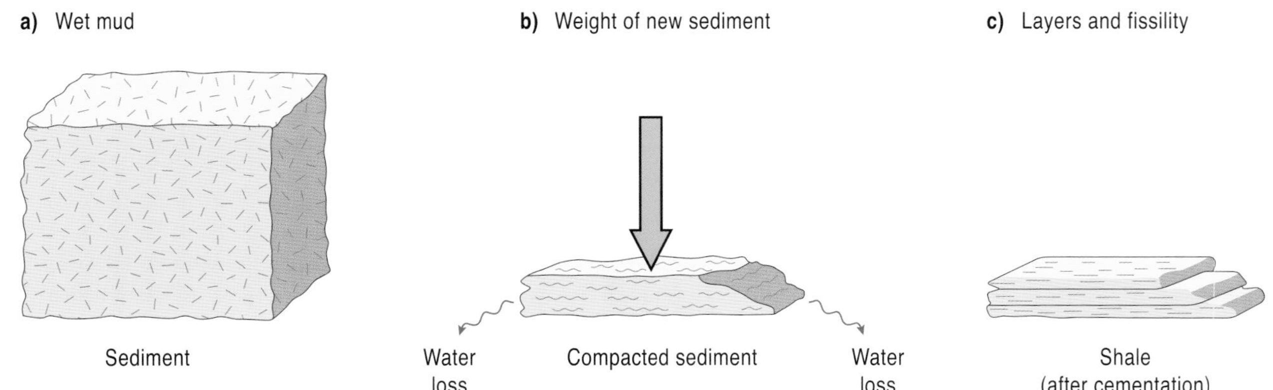

Figure 2.11 At the time of deposition, muds can constitute up to 90% water and undergo rapid dewatering during burial leading to a mudrock. **A**) Wet mud at time of deposition. **B**) Silt and clay particles tend to align with loss of water, and overall pore space decreases during burial with loading from new sediment on top. **C**) Through cementation a mudrock is formed that tends to be smooth to the touch, and when visibly layered and fissile is called a shale.

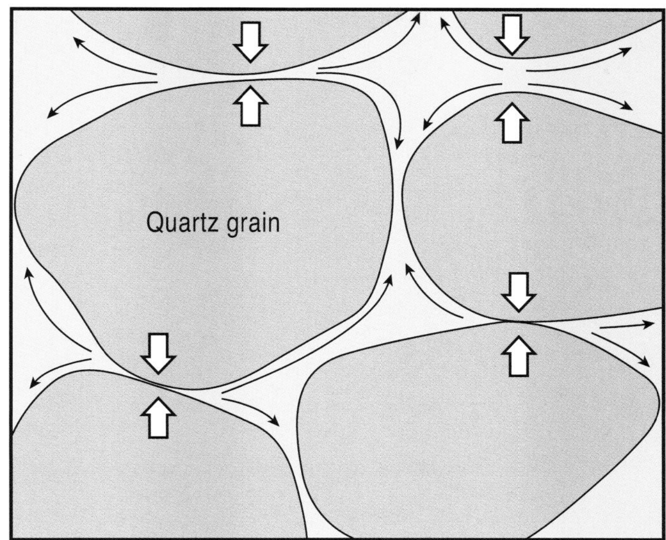

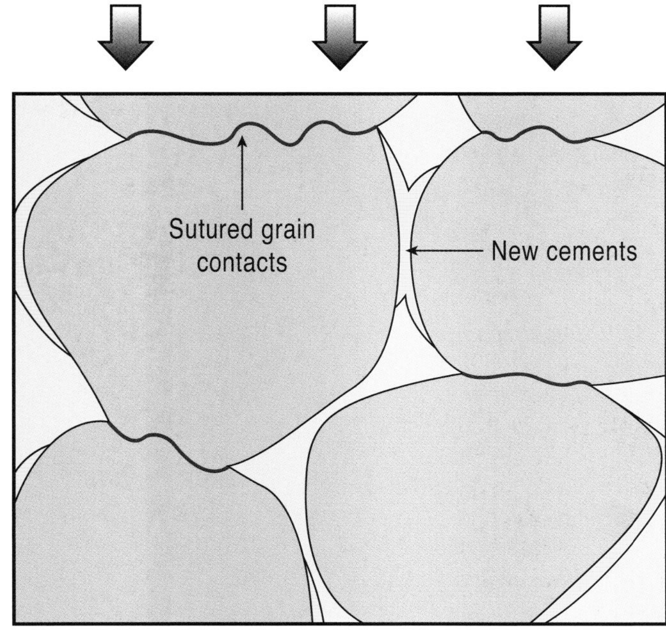

Figure 2.12 As a sediment is progressively buried under new layers, it starts to experience chemical and physical changes. It is common for grains to experience greater concentrations of stress at points of contact, leading to pressure dissolution, where grains start to dissolve and to develop sutured contacts. Pressure dissolution is common in many quartz-arenites.

of chemical reactions between the rock and new fluids passing through the rock, and as the result of increasing pressure and temperature. These reactions may dissolve previously formed cements and/or form new cements and minerals (Fig. 2.12). Commonly, in response to changes in pressure and temperature mineral transformations occur that alter the mineralogy of the sedimentary rock. This commonly occurs amongst clay minerals and in carbonate rocks, where aragonite undergoes transformation to calcite.

In mudrocks, not only are clay mineral transformations a common occurrence, but also simultaneous alignment of clay minerals leads to the formation of a visibly layered and fissile fabric, producing shale (Fig. 2.11).

Appreciating diagenetic processes is important for several reasons. They firstly can modify the starting sediment, both in terms of texture and composition, and even remove primary sedimentary structures. Secondly, diagenesis can affect the porosity and permeability of a sedimentary rock, these being important physical properties to consider in the exploration for oil, gas or water reservoirs (Figs 2.13 & 2.14; see also Table 6.1).

Figure 2.13 The diagenesis of an oolitic sediment to form oolitic limestone.

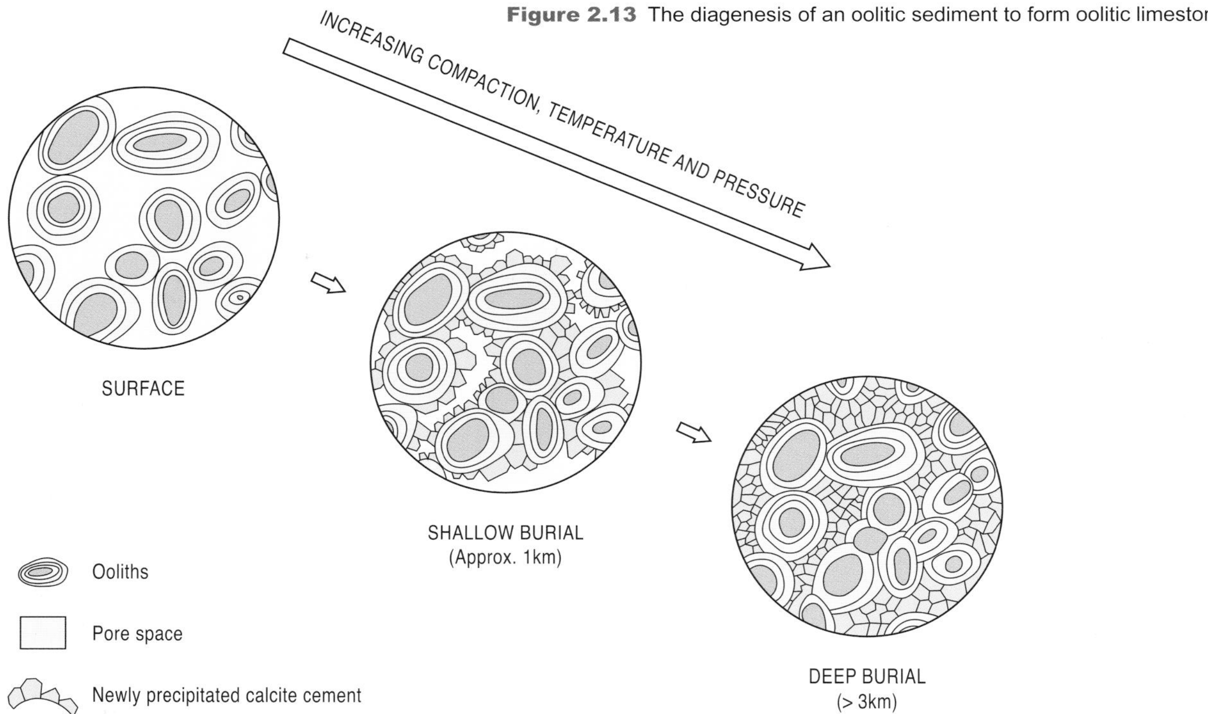

Figure 2.14 Even at great depths (>4 km) beneath the Earth's surface, sediments can preserve anomalously high porosity (>30%). In such occurrences the compaction processes must have been slowed, and this is often attributed to a combined array of diagenetic processes. The photomicrograph illustrates a Triassic age fine-grained sandstone from the high-pressure, high-temperature hydrocarbon region in the Central North Sea, UK. The blue colour is a dye added to the rock to identify porosity.

3 Sedimentary structures

Deciphering the visible marks left by natural processes on and in sedimentary rocks is a fascinating part of a sedimentologist's work. The analogy of a forensic scientist can be drawn from the detailed investigations undertaken at a crime scene, where even the smallest detail is collected as evidence to piece together the sequence of events. Sedimentary textures such as sorting and rounding of grains can give clues to depositional processes for the sedimentologist. Windblown sand grains tend to be well rounded and well sorted, but textures alone tend to be insufficient to determine depositional environments. Other sedimentary rock properties need to be considered. The marks, traces and sediment disturbances preserved in the sedimentary strata are called **sedimentary structures** and are a reliable way for the sedimentologist to determine the geological history of a sedimentary sequence. To make the most of sedimentary structures as a tool, both their potential for preservation and limitations of interpretation must be known. Among the many structures that can be produced in sediments, only a few are bound to be preserved in the rock record. Most structures are cancelled out by the same sedimentary process that created the feature, or by prolonged exposure to erosion and weathering. Any newly deposited sediment is subject to reworking and remobilization, for example by storm events. Only once the sediment has been buried can it preserve its structures. How many and which structures will be found in the stratigraphic record depends not on a single event but on the whole geological history of the sediment.

Perhaps the biggest challenge in identifying sedimentary structures is that no two structures are the same, even those produced by the same sedimentary process. Each structure has a certain degree of individuality.

Which way is up?

Knowledge of the fundamental principles of stratigraphy (as in chapter 1) is especially important for correctly reading and interpreting sedimentary strata. The **principle of superposition**, where sedimentary layers are deposited in a time sequence, with the oldest on the bottom and the youngest on the top, uses **way-up criteria** to establish whether **beds** are in normal position or upside down because of tectonic movements (Fig. 3.1). Every time a sedimentologist observes inclined strata we ask ourselves, are these rocks the correct way up or overturned? The answer in many cases can be provided by sedimentary structures. We identify the sedimentary structures, their orientation and shape to give the correct orientation of the sedimentary strata.

This technique is particularly important in areas affected by compressional tectonics (as in mountain belts) and where there is a lack of other indications of the relative ages of beds within the sequence, such as in the Precambrian (rocks older than 540 million years).

One of the most useful way-up criteria are **geopetal structures**. These are, in effect, a fossil 'spirit level' that clearly identifies the former horizontal of the beds, and importantly, their correct orientation. These represent the double filling of a cavity or void in a sedimentary rock (Fig. 3.2). The first internal sediment is horizontally layered and then later on remaining space is occupied by a chemical cement (often calcite or quartz), whose crystals seal the structure. Geopetal structures are particularly useful in poorly bedded rocks like carbonate reefs where internal cavities are common. Examples of geopetal structures have also been recognized in the 400-million-year-old Rhynie chert of Scotland (Fig. 3.3).

What sedimentary structures tell us about the rocks

Physical and biological processes operating in sedimentary environments are responsible for a variety of structures. It is the array of different structures that allows sedimentary facies to be assigned (chapter 1). These are the building blocks of sedimentology. Sedimentary structures can be subdivided into a number of different types.

Figure 3.1 Inclined beds of Silurian Aberystwyth Grits, UK. It is always important to determine the way-up of the bedded strata. © Denis Bates, Aberystwyth University, Wales, UK.

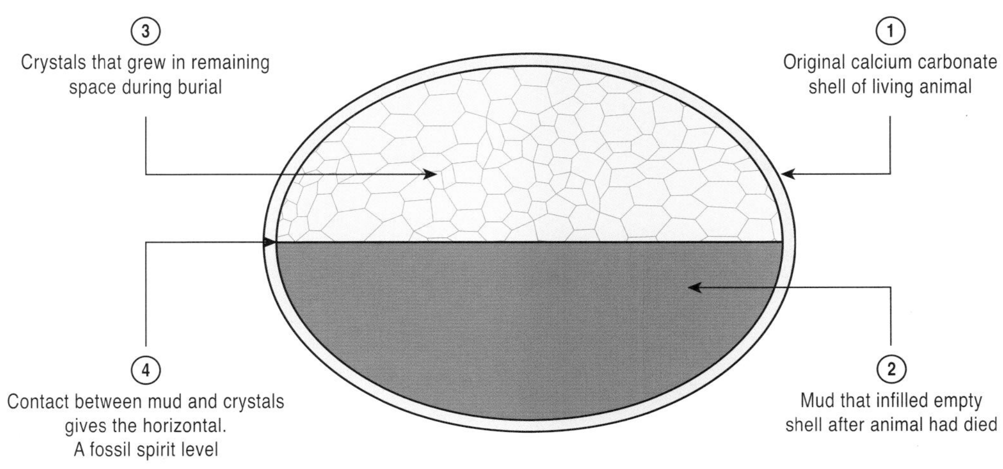

Figure 3.2 Formation of geopetal structures. They are used as an indication of the palaeohorizontal level.

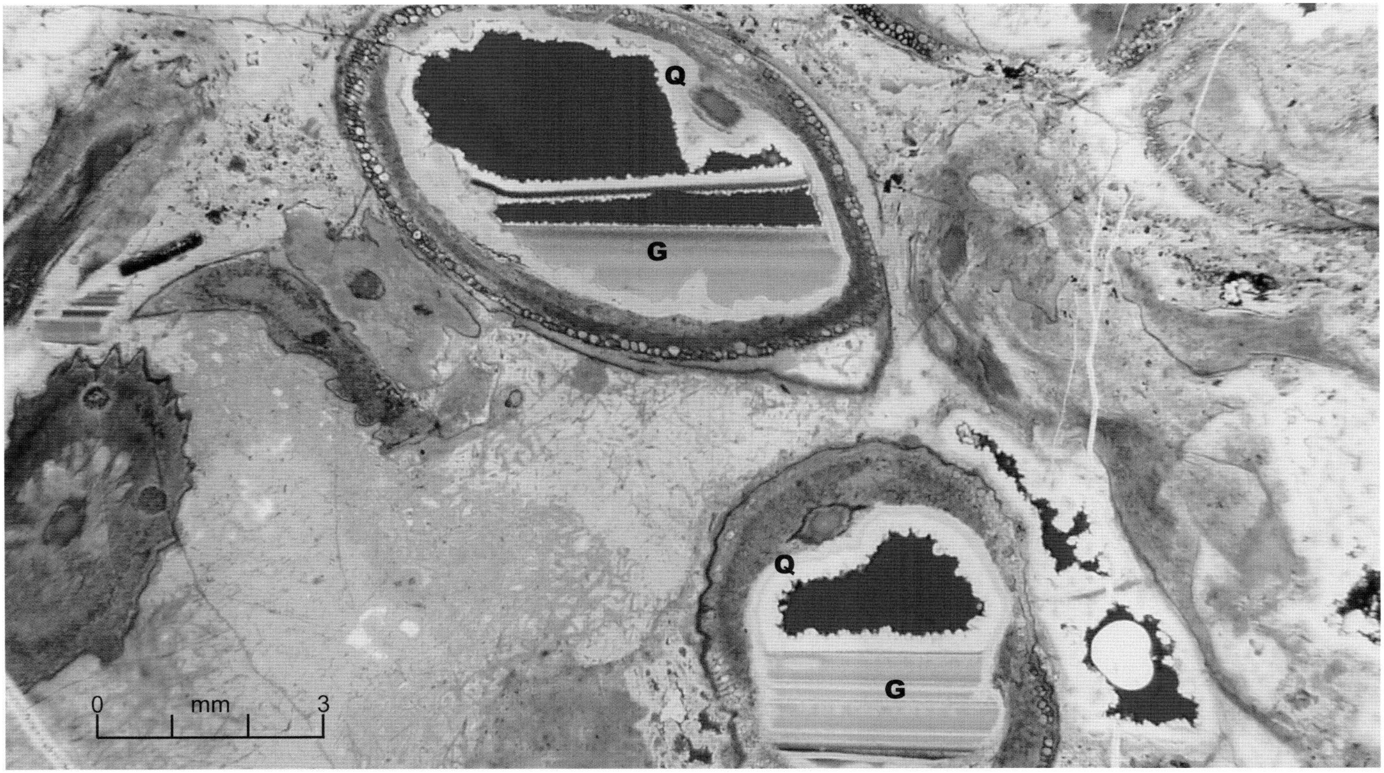

Figure 3.3 Photomicrograph of Devonian Rhynie chert from Aberdeenshire, Scotland. Cross-sections through plant stems of the plant *Aglaophyton major*. Geopetal layers of very fine sediment within the plant stems indicate the 'way up'. The blue colour in this thin section is stained epoxy resin infilling pore space (diameter of plant stem = 3 mm). Geopetal layers (**G**) denoting the image is the correct 'way-up'. Later generation of quartz cement (**Q**) lining the remaining void space after the geopetal fill. © Nigel Trewin, Professor Emeritus, Aberdeen University, UK.

Beds and laminations

Beds or strata are one of the most distinctive and common of all sedimentary structures. Beds (thicker than 1 cm) and laminae (thinner than 1 cm) are usually separated from one another based on changes in grain-size, colour, composition, texture, or a combination of features (Table 3.1). Layering of some kind within beds is present in all sedimentary rocks, but is frequently best illustrated where there exists a large grain-size variation (e.g. between beds of sandstone and mudstone; Fig. 3.4).

Table 3.1 Terminology used by sedimentologists for the thickness of beds and laminae.

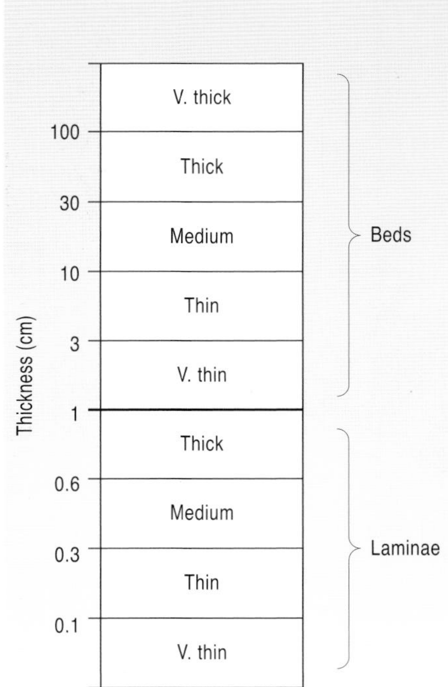

Figure 3.4 The Aberystwyth Grits comprise several hundred metres thickness of mudstone and sandstone beds laid down in the marine Welsh Basin during the Silurian. They are interpreted as a sequence of turbidites (*see* Fig. 4.15). © Denis Bates, Aberystwyth University, Wales, UK.

Erosional structures

These tend to be formed by the action of erosive currents with the removal of sediment to form the structure. The most familiar are **flute casts** (also known as **scour marks**) that occur on the underside of many beds deposited by sediment-laden flows and storm currents (Fig. 3.5).

Flute casts are perhaps the most distinctive of all erosional structures, with a heel-shape consisting of a bulbous upstream end that flares downstream and merges with bedding. The formation of flutes is attributed to localized erosion of sand-laden currents passing over cohesive muds. This is a common process in submarine environments, where sediment-laden flows can readily occur. Measuring the long axis of the flute cast gives the direction of flow, with the tapered end pointing down-current and the steep end up-current (Table 3.2).

In comparison **tool marks** are formed when an object such as a

Figure 3.5 Flute casts on the base of a sandstone bed in the Silurian Aberystwyth Grits, UK. Measuring the long axis of the flute cast gives the direction of flow, with the tapered end pointing toward the flow and the steep end up-current. The concavity of the flute cast also points stratigraphically up. Note geological hammer on right of image for scale. © Denis Bates, Aberystwyth University, Wales, UK.

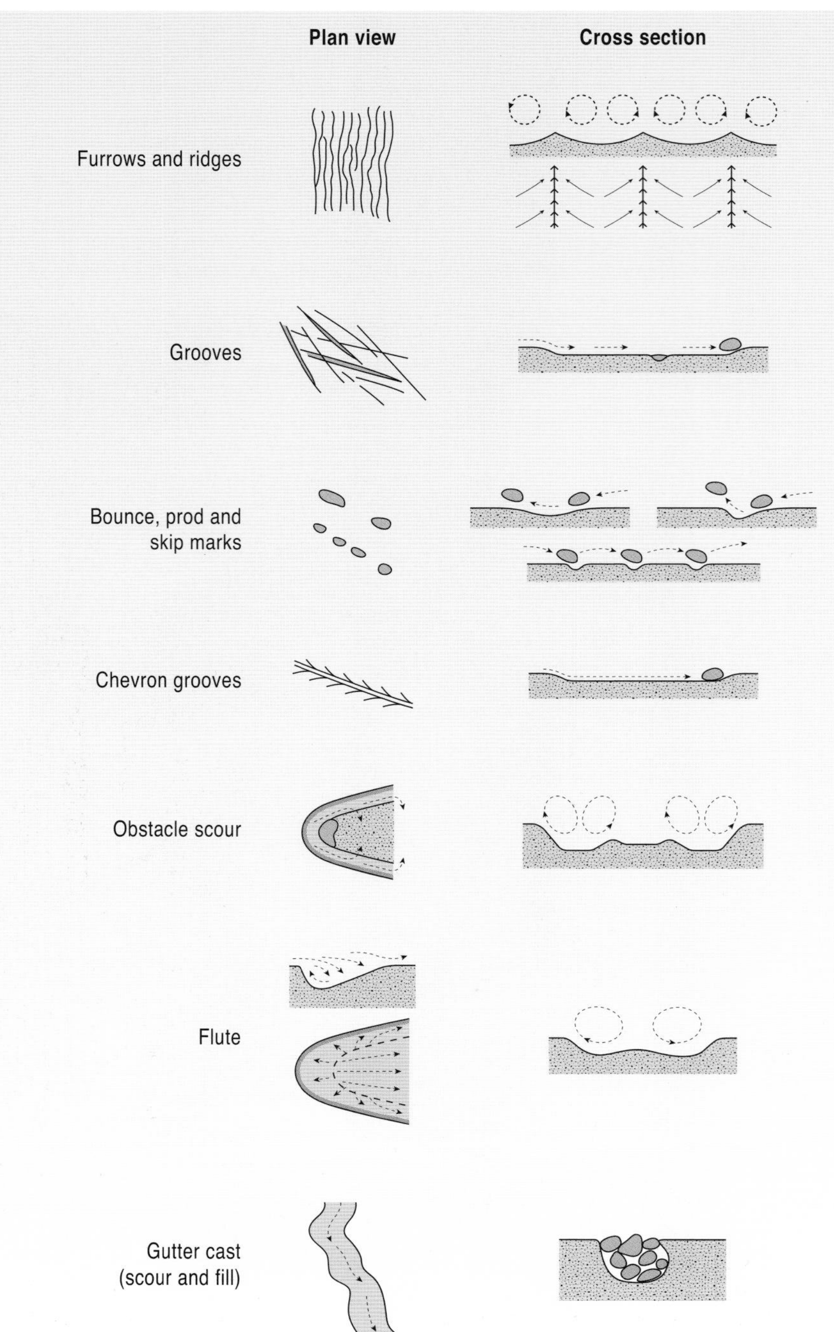

Table 3.2 Commonly encountered erosional structures formed by the erosive action of currents. They usually occur as casts on the underside of beds, or depressions on bed surfaces.

pebble or fossil is carried by a moving current and impacts or scrapes along a sediment surface (Table 3.2). Discontinuous tool marks include prod, bounce, skip and roll markings and tend to be associated with highly turbulent flows. More continuous tool marks include grooves and chevrons on the undersides of sandstone beds. Grooves can occur as single linear mark or as groups of markings (Table 3.2). It is considered that grooves form in a similar way to other tool marks, with a fossil or pebble that was carried along by a current gouging a groove into the underlying sediment, with best results preserved in soft, muddy sediments. Groove marks can occur in a wide variety of sedimentary environments where sand and pebbles are regularly moved, such as on shallow-marine shelves with storm currents or when a river is in flood and breaks its banks and spreads out over the flood plain.

Perhaps some of the rarer examples of sedimentary erosional structures are created by raindrops, which can leave imprints in fine-grained sediments (Fig. 3.6). The use of ancient raindrops was first recognized in 1851 by the pioneering British geologist Charles Lyell, who suggested that raindrop indentations could be used to estimate ancient air pressure and as a proxy for how this may have changed through geological time.

Figure 3.6 Fossil raindrop impressions from the Pliocene, Sorbas Basin, Southern Spain. They are characterized by small craterlike pits with slightly raised edges that are the result of the impact of rain on soft sediment surfaces. In this case the raindrop impressions are preserved in fine-grained lagoonal muds.

Depositional structures

Depositional structures form during the deposition of sediment, usually on the bedding surface. They tend to be the most common of all sedimentary structures preserved, and occur as a wide variety of features.

Bedforms are features formed by the interaction of either water or wind and sediment on a bed surface. Examples include ripples in sand found in a flowing stream. The recognition of sedimentary structures produced by bedforms provides valuable information about the speed of the current, the direction of sediment movement and the flow depth. One of the most common and easily recognizable bedforms is that of current ripples. **Current ripples** are characteristically asymmetrical in cross-section, with the down-current slope steeper than the up-current one (Fig. 3.7a). In plan view, the ripples tend to be fairly straight-crested (Fig. 3.8).

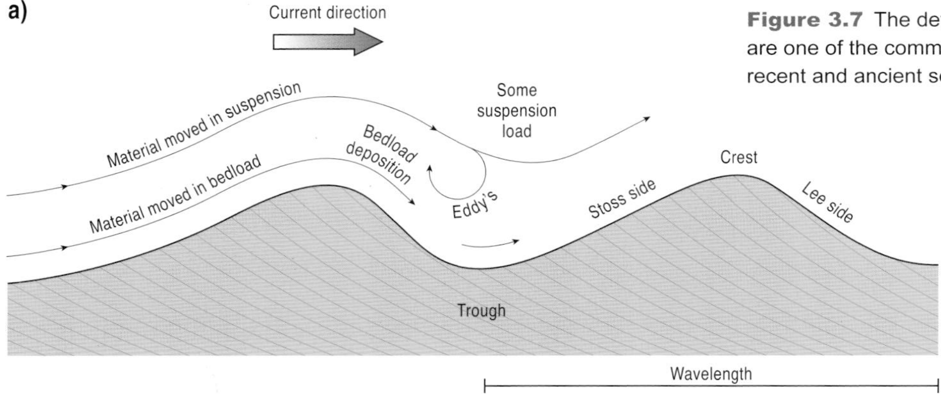

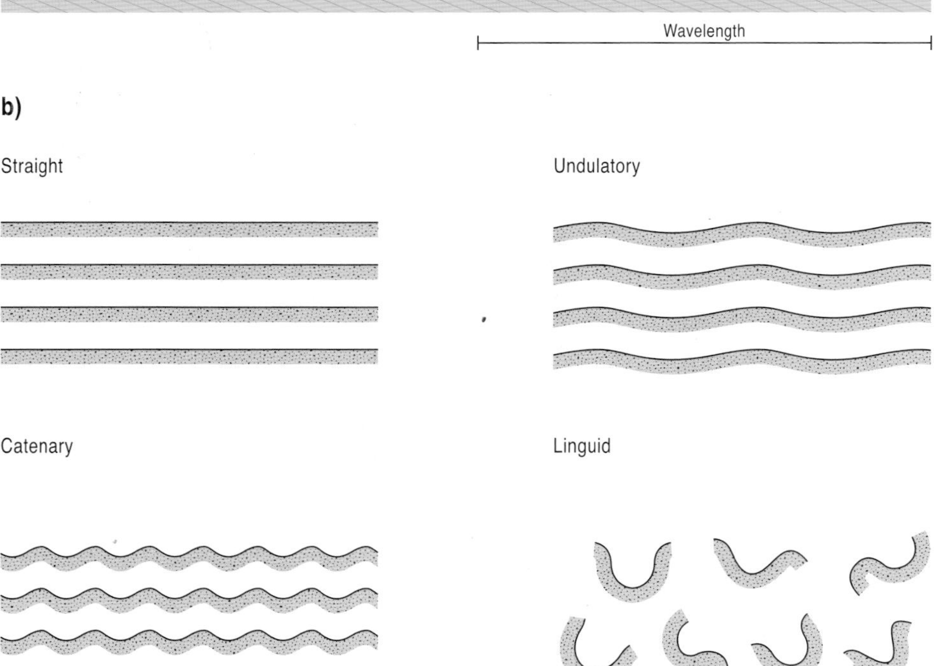

Figure 3.7 The development of current ripples. Ripple marks are one of the commonest features of sedimentary rocks, both in recent and ancient sediments.

Current ripples are about 2–5 cm in height and with wavelengths of up to ~40 cm. These ripples are not static features: they migrate along the bed in the direction in which the current is flowing (Fig. 3.7a). Sediment is eroded from the stoss side (shallow upstream slope) and carried to the crest, from which it avalanches down the lee side (steep downstream slope). Suspension load is carried over the top of the ripple and some is deposited on the lee side, where eddies develop in the flow. In this way, a series of laminae build up parallel to the surface of the lee slope, inclined in the direction of current flow. Successive laminae build up as the bedform migrates downstream and gives rise to cross-lamination, or more commonly termed **cross-stratification** (Fig. 3.7a).

As a flow current increases the ripple crests steadily become more sinuous (Fig. 3.7b) and eventually break up into curve-crested segments, creating complex linguoid (tongue-shaped) ripples (Fig. 3.9). Increasing the current speed further produces larger bedforms known as **dunes**. These are up to 1 m high and with wavelengths of ~0.5 m to ~10 m. If the flow speed is increased even further, then dunes are washed out and sediment sweeps over a flat surface, creating planar stratification (Fig. 3.10). At speeds higher still, the flat bed changes into low relief, undulating mounds known as antidunes (Fig. 3.10). These bedforms occurring at higher flow velocities are rarely preserved in the geological record because they are usually reworked as soon as the flow speed decreases.

Interpreting sedimentary structures such as ripples and associated

Figure 3.8 Straight-crested symmetrical ripples. Carboniferous sandstone, Cullercoats, Northumberland, UK.

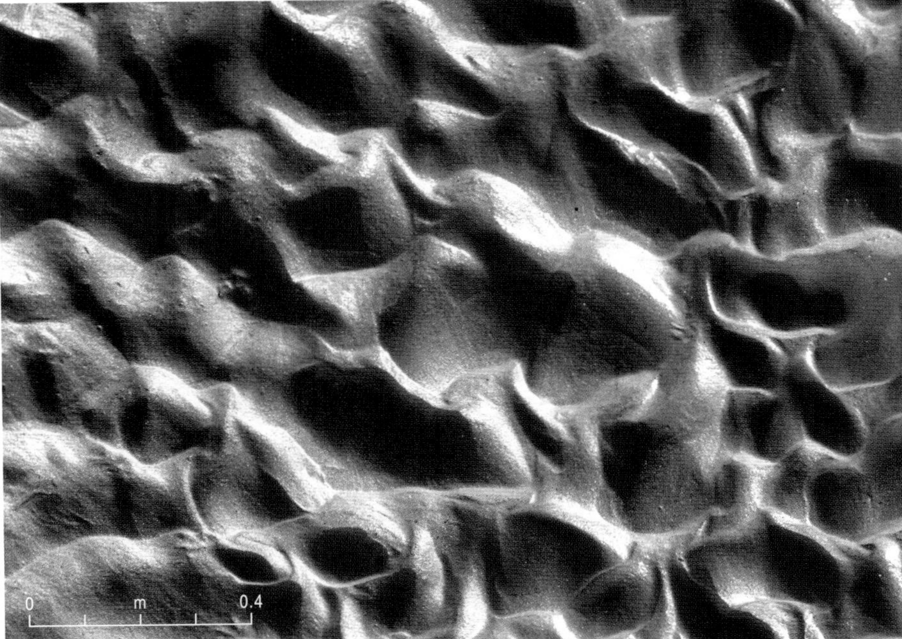

Figure 3.9 Linguoidal (tongue-shaped) ripples, the Wash, North Norfolk coastline, UK.

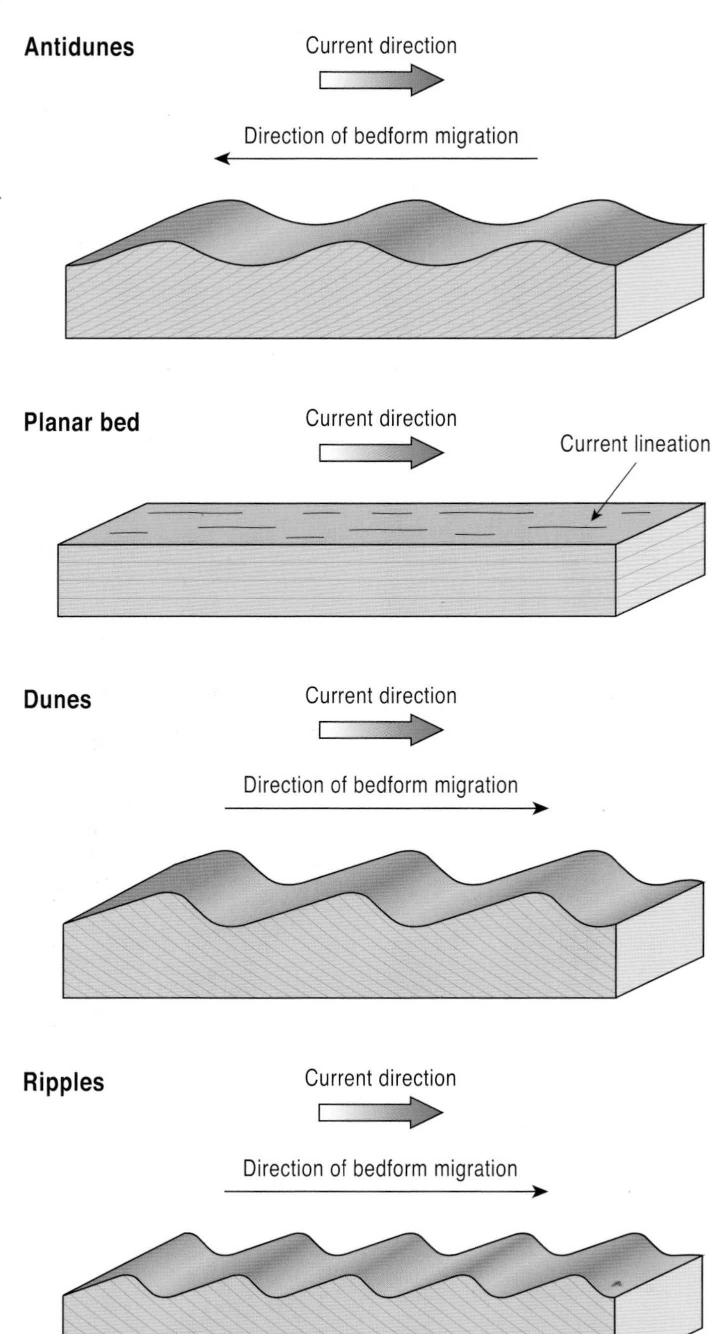

Figure 3.10 Sedimentary bedforms. Bedforms are often characteristic of the flow parameters, and may be used to infer flow depth and velocity.

cross-stratification in the field is rarely a simple task. Variations occur due to differences in current directions, shapes of bedform crests and flow velocities. However, there are two basic types of cross-stratification as shown in Figure 3.11. Planar cross-stratification is produced by two-dimensional bedforms (those with straight crests) and trough cross-stratification is produced by three-dimensional bedforms (those with curved and sinuous crests).

Where mixtures of sand deposited by ripple migration and mud deposited from suspension fallout occur, then distinctive sedimentary structures may result, depending on the relative proportions of sand and mud (Fig. 3.12). Flaser, wavy and lenticular bedding are all formed in environments of deposition where alternating periods of moving and slack water occur, as in tidally dominated marine settings where there is daily variation in flow regime. Mud drapes over bedforms, such as ripples, greatly enhance their chances of preservation.

The oscillatory motion of waves in shallow marine and intertidal settings can produce **wave ripples** (Fig. 3.13). These ripples are in contrast to current ripples, as they are characterized by a symmetrical profile and usually continuous straight crests. Wave-generated oscillatory flows produced by the passage of storms generating large storm waves produce a large-scale undulatory bedform known

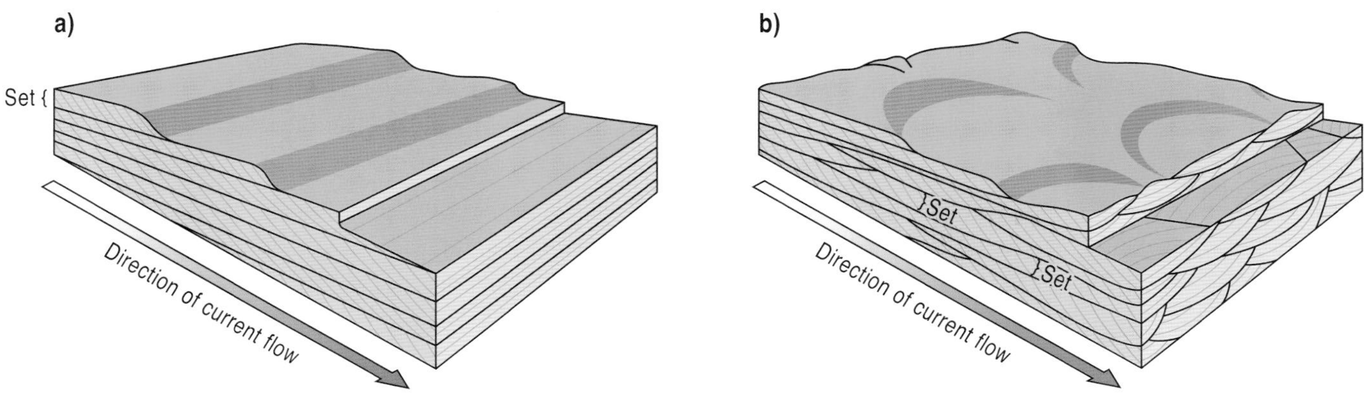

Figure 3.11 Cross-bedding is formed by the downstream migration of bedforms such as ripples or dunes. The downstream migration of ripples forms planar cross-bedding (A) and the migration of dunes produces trough cross-bedding (**B**).

Figure 3.12 Flaser and wavy bedding: sedimentary structures characterized by the alternation of rippled sand and mud layers. These structures are often considered to form mostly in tidally influenced environments. Carboniferous, Howick, Northumberland, UK.

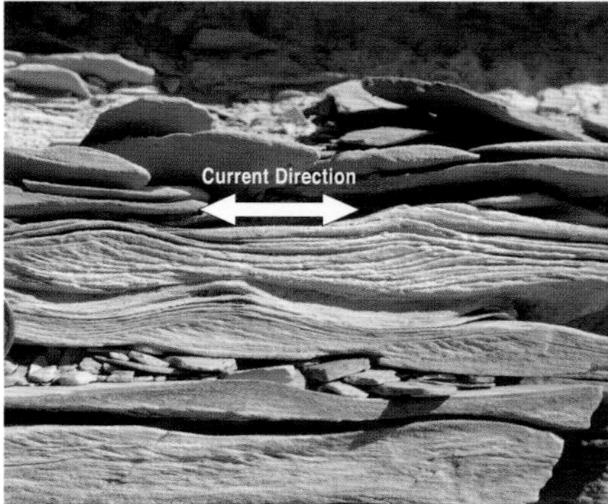

Figure 3.13 Ripple cross-laminated sandstone, showing bi-directional cross laminae indicative of a wave origin. Entrada Formation, Jurassic, San Rafael Swell, Utah, USA.

Figure 3.14 Large-scale aeolian cross-bedding. Jurassic age Navajo Sandstone, Utah, USA.

as **hummocky cross-stratification (HCS)**. HCS takes on the form of hummocks and swales that are circular to elliptical, with long wavelengths (1–5 m) but with low height (a few tens of centimetres). With the passing of the storm conditions, normal fair-weather conditions are resumed and HCS is usually overlain by wave-rippled sands and extensive **bioturbation**.

Bedforms formed in deserts are perhaps the most awe-inspiring products of sediment transportation. Truly gigantic bedforms can exist; for example, in the Saharan sand seas dunes can form up to 100 m in height.

Wind ripples tend to be very similar to those of current ripples but generally with better sorting, rounded grains and straight-crested and asymmetric in form. However, it is the larger-scale bedforms, such as dunes and draas, which have wavelengths up to several kilometres, that are a more common sight (Fig. 3.14).

Graded beds

Some form of systematic change in grain size from the base of a bed to the top is very common in sedimentary strata. Most commonly this takes the form of normal grading, with coarser sediments at the base, which grade upward into progressively finer ones (Fig. 3.15). Normally graded beds generally represent depositional environments that decrease in transport energy over time, but also form during rapid deposition. Inverse grading, in comparison, is where grain size increases up through a bed (Fig. 3.15). This type of grading is relatively uncommon but can occur in debris flows and is also observed in aeolian ripples.

Post-depositional features

Many sedimentary structures can form immediately following deposition, where sediments can be very poorly consolidated and subject to wet-sediment deformation and larger scale slumping. These processes vary in scale from the movement of large volumes of sediment through slumping and sliding through to shrinkage at the sediment surface from drying out.

Slumps are common occurrences in the geological rock record and at the present day. They occur as the result of gravitational instabilities in piles of sediment in both continental and marine settings. If a slope is subjected to an earthquake or sudden

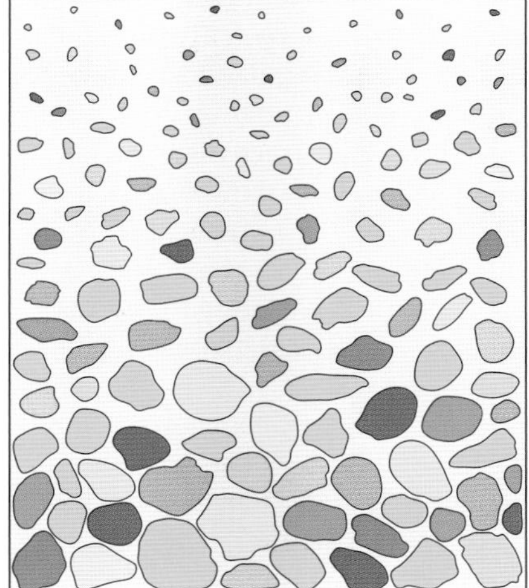

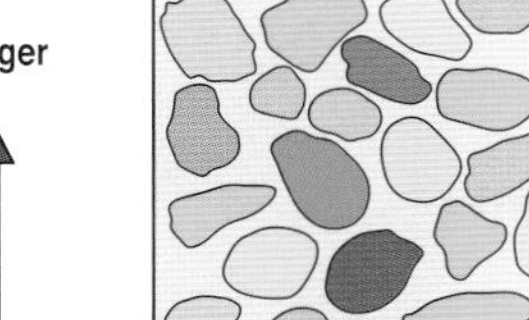

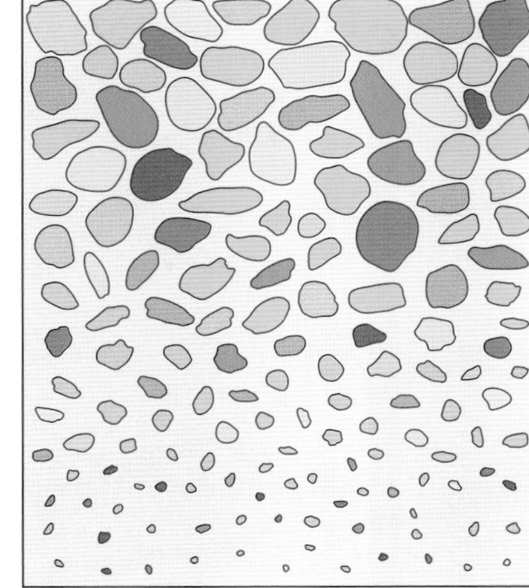

Figure 3.15 Types of graded bedding commonly found in sedimentary rocks.

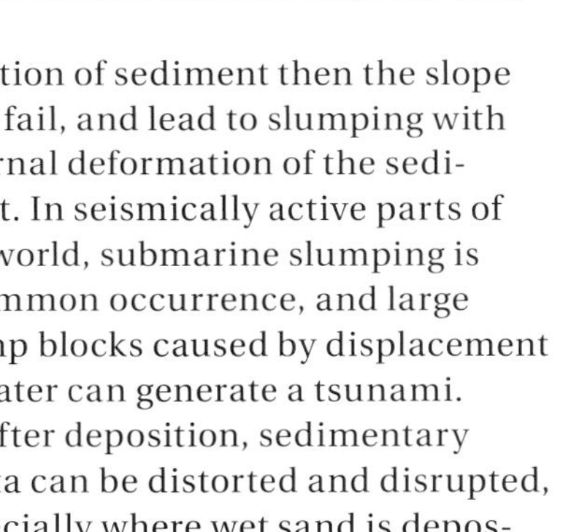

addition of sediment then the slope may fail, and lead to slumping with internal deformation of the sediment. In seismically active parts of the world, submarine slumping is a common occurrence, and large slump blocks caused by displacement of water can generate a tsunami.

After deposition, sedimentary strata can be distorted and disrupted, especially where wet sand is deposited on top of wet mud. The weight of the overlying sediment upon the wet, unlithified mud causes the less dense mud to rise and penetrate into the overlying sand bed. This can give rise to flame structures or broader down-bulges of sand into mud in the form of load casts (Fig. 3.16). Mudcracks or desiccation cracks can be found with wet sediments and are common in fine-grained sediments where the mud dries and contracts. They are indicative of sub-aerial exposure at the land surface and can be used as way-up criteria (Fig. 3.17).

Figure 3.16 Flame and load structures at the base of a turbidity current from the Late Miocene, Tabernas Basin, SE Spain.

Figure 3.17 (**A**) Modern mud cracks. (**B**) Cross-sectional view of a mud or desiccation crack with a fine-grained sandstone infilling the crack. Carboniferous Mabou Group, Nova Scotia, Canada. Photo by Michael C. Rygel.

Biogenic structures

The action of plants and animals provides many structures preserved in sediments. These include the irregular disruption of the sediment (bioturbation), discrete organized markings (trace fossils) and direct biogenic growth structures (stromatolites) (Fig. 3.18 a, b). Trace fossils are biogenic structures that record the activity of organisms such as burrowing, walking and feeding. They are particularly useful to the sedimentologist as they are usually preserved *in situ* and are an accurate and more reliable indicator of an ancient sedimentary environment than **body fossils** (e.g. ammonites, trilobites, etc.). Although trace fossils (**ichnology**) can be found in all sedimentary environments they tend to be best preserved in marine settings. The different types of trace fossils preserved are classified based on inferred behaviour such as feeding, dwelling and locomotion of the organism that formed the trace. This provides greater insight into the sedimentary depositional setting (Fig. 3.19).

Perhaps where trace fossils are most valuable is in their application to palaeoenvironmental interpretations. The behaviour of an organism is linked to environmental conditions such as water depth, rates of sediment deposition, oxygen abundance and salinity levels. As a result, different trace fossils form distinct groupings that may occur in shallow water conditions with high energy and shifting sands, or perhaps in deeper water conditions where quiet, moderately oxygenated conditions exist. Each group of distinct trace fossils is known as an **ichnofacies** that can be particularly useful for sequence stratigraphy and attaining more detailed palaeoenvironmental information.

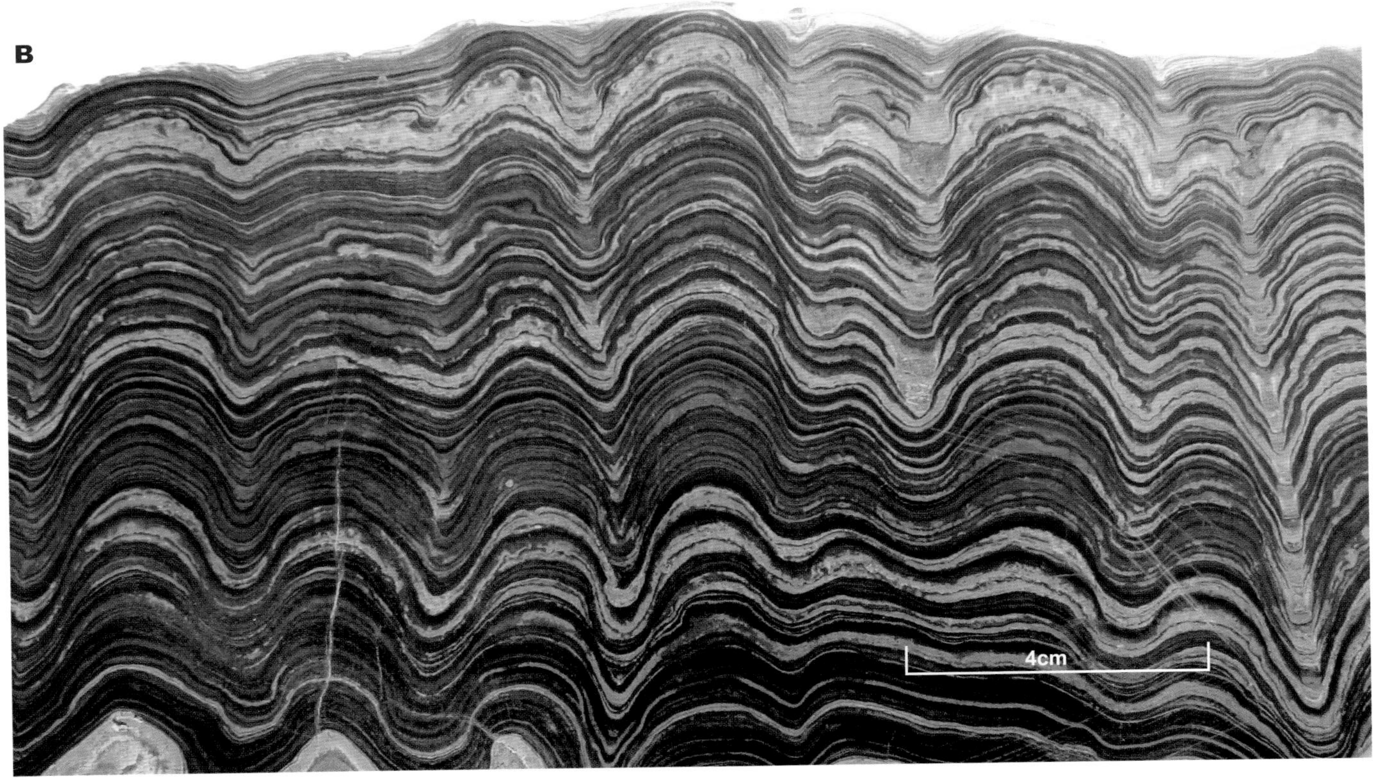

Figure 3.18 Types of biogenic sedimentary structures. (**A**) Trace fossils in a shallow-marine sandstone where individual burrows are vertical or inclined and cross-cut the horizontal stratification. Carboniferous, Whitley Bay, Newcastle, UK. (**B**) Stromatolites are layered biochemical accretionary structures formed in shallow water by the trapping, binding and cementation of sedimentary grains by biofilms (microbial mats) of micro-organisms, especially cyanobacteria. Polished section of a Precambrian stromatolite, Lake Uguyi, Bolivia.

B

4cm

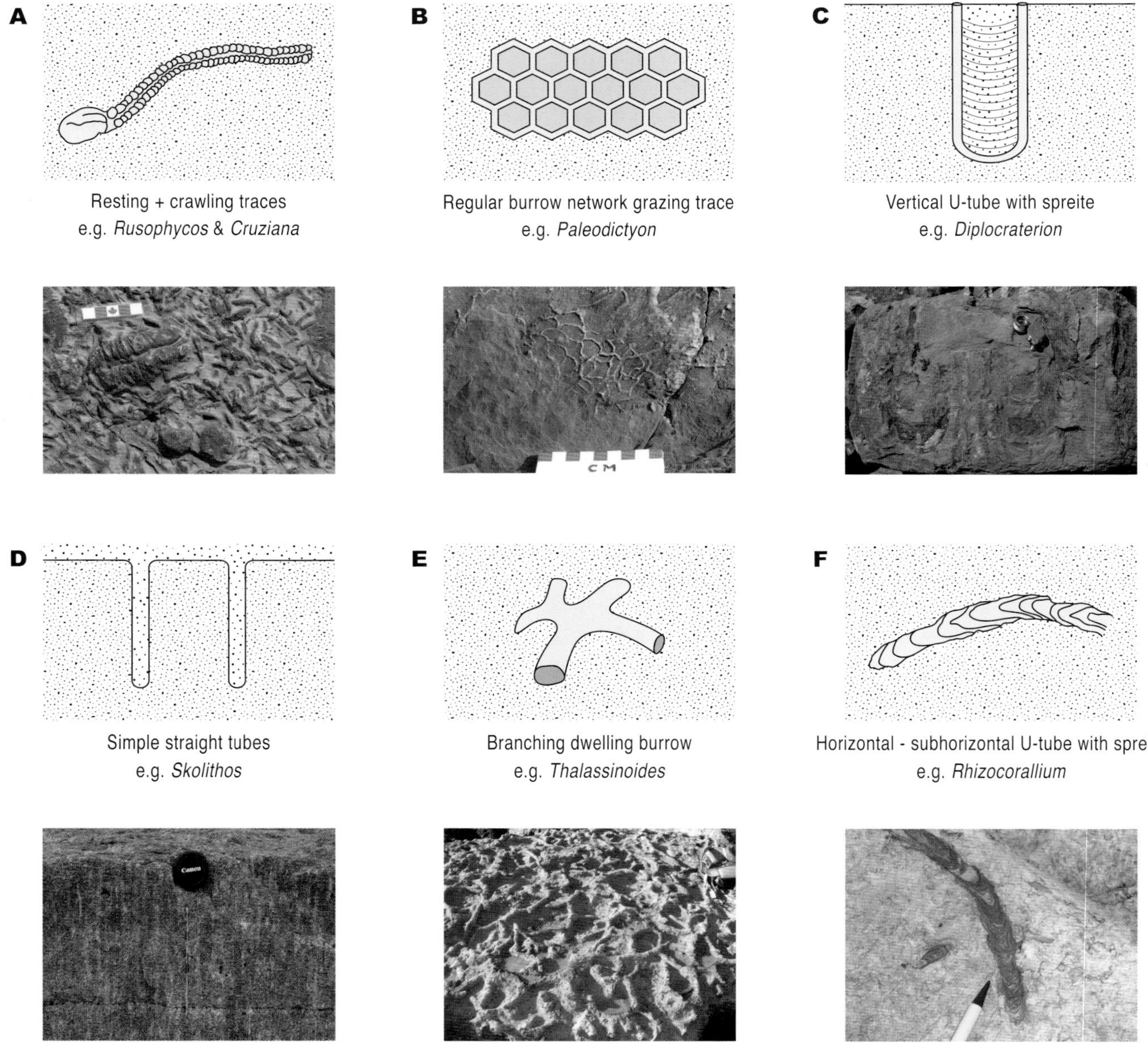

Figure 3.19 Trace fossils (**A**) Resting and crawling traces. (**B**) Network grazing traces. (**C**) An 'escape burrow' where the organism moves upward during high-sedimentation rates, creating **spreite** below it, in order to prevent the burrow entrance from being buried. (**D**) Vertical simple burrows. Skolithos burrows have rarely been described from carbonates. (**E**) Branching, dwelling burrows, thought to have been produced by burrowing crustaceans. (**F**) Sub-horizontal burrows that may be dwelling and/or feeding burrows. These burrows can be very large, over a metre long in sediments that show good preservation, e.g., Lower Jurassic Cleveland Ironstone, North Yorkshire, UK. ©Liam Herringshaw.

Dinosaur footprints

Dinosaurs are a fascination for all ages, and perhaps one of the most revealing aspects of dinosaurs are their preserved footprints (trace fossil) in sedimentary rocks. Dinosaur footprints reveal direct information about their locomotion, social behaviour and general morphology. They are usually preserved on bedding surfaces and their preservation is strongly dependent on the type of sediment the dinosaur was walking over. However, equally the sediments can provide extra information as to the palaeoenvironmental setting where the dinosaur once lived. Where several footprints created by a single dinosaur are found on a sedimentary bedding surface, a trackway can be preserved, which can indicate if the dinosaur was bipedal or quadrupedal. Detailed analysis of trackways can be used to estimate the speed of a dinosaur and even determine if the dinosaur was injured.

The social behaviour of dinosaurs is uniquely identified from trackways, where parallel tracks can indicate herding or migratory behaviour. Similarly associated with such migratory tracks, those of predators can be found, which have assisted in determining predator–prey relationships.

The weird and wacky!

There are many sedimentary structures that perhaps do not fit easily into any sort of classification or have long been misidentified as to the actual processes that formed them. One such example is **septarian nodules**, recognized as concretions or nodules containing cracks. The word comes from the Latin word septum, 'partition', and refers to the cracks/separations in this kind of rock. Cracks are highly variable in shape and volume, as well as the degree of shrinkage they indicate (Fig. 3.20). Although it has commonly been assumed that concretions grew incrementally from the inside outwards, the fact that radially oriented cracks taper towards the margins of septarian concretions is taken as evidence that in these cases the periphery was stiffer while the inside was softer, presumably due to a gradient in the amount of cement precipitated and hardening of the sediment. The process that created the septaria that characterize septarian concretions remains a mystery. A number of mechanisms have been suggested, such as the dehydration of clay-rich, gel-rich, or organic-rich cores; shrinkage of the concretion's centre; expansion of gases produced by the decay of organic matter; and brittle fracturing or shrinkage of the concretion interior by either earthquakes or compaction.

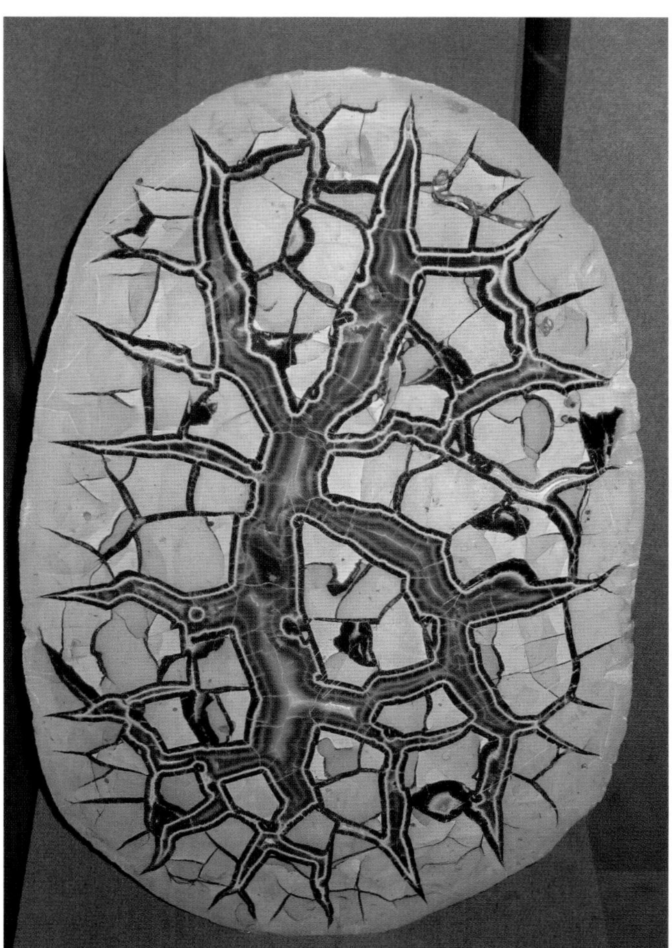

Figure 3.20 A septarian nodule on display as part of the Pinch Collection at the Canadian Museum of Nature in Ottawa, Ontario. Photo by Keith Pomakis – Wikipedia image with common use.

4 The sedimentary environment

Every sedimentary rock bears the imprints of the natural forces that shaped it. By examining the layers of sedimentary rocks that accumulate over geological time, the sedimentologist can decipher the Earth's past environments. In most cases the environments associated with particular sedimentary rock types can be matched to modern day environments (e.g. a river or coral reef). However, the further back in geological time sediments were deposited, the more likely it is that direct modern analogues are not available. Sedimentologists interpret and compare rocks they examine with modern sedimentary environments using a combination of approaches, as described in chapters 1, 2, and 3. It is vital when interpreting sedimentary rocks and successions that the rocks are considered at all scales from the individual grain up to the large-scale basin fill (Fig. 4.1). Frequently a sediment's composition,

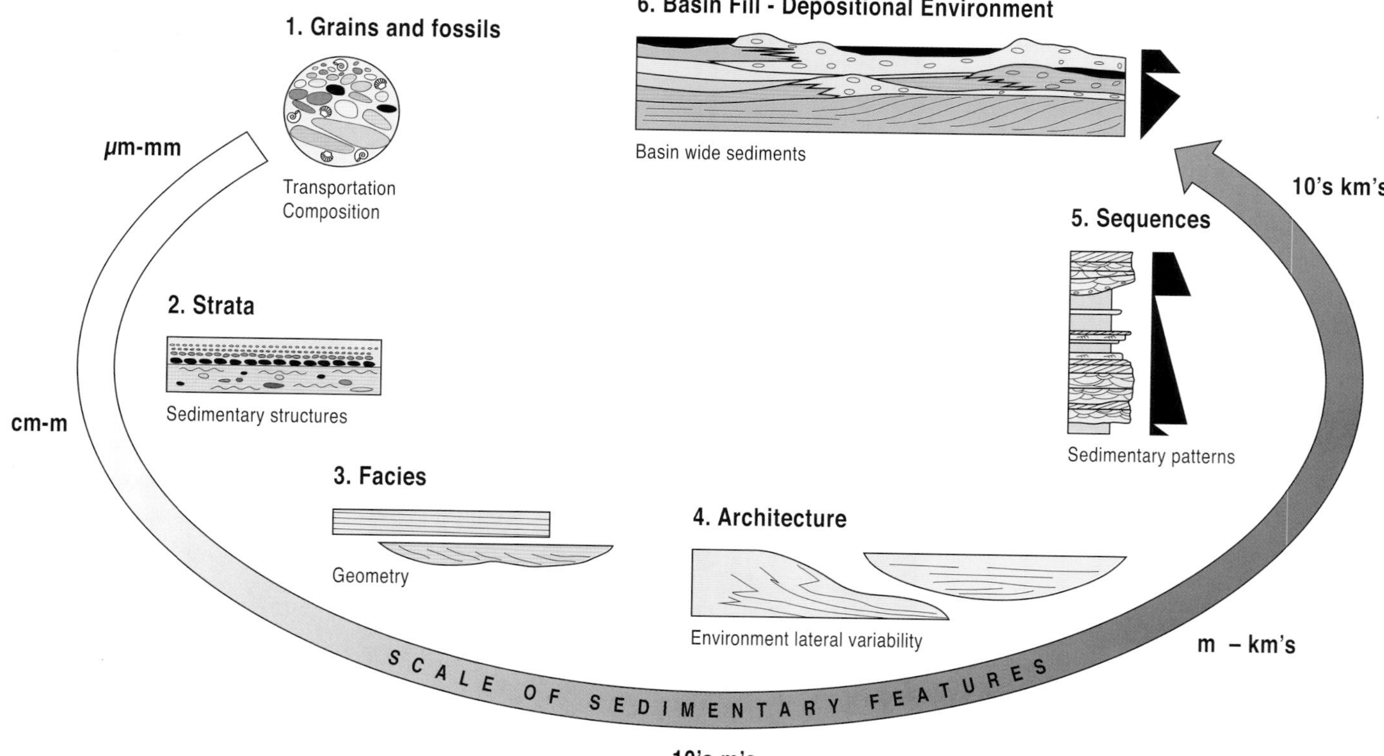

Figure 4.1 From grain to basin scale. It is important when interpreting sedimentary successions that all scales of processes are considered. Adapted from Heinz, J. and Aigner, T. (2003) Three-dimensional GPR analysis of various Quaternary gravel-bed braided river deposits (southwestern Germany). In: Ground Penetrating Radar in Sediments (Eds C. Bristow and H. Jol), Special Publications, 211, pp. 99–110. Geological Society, London. Also modified with reference to: http://www.sepmstrata.org/page.aspx?pageid=410 [Accessed 2014].

a sedimentary structure or a fossil, are not unique to a specific sedimentary environment, and as much sedimentological information as possible must be gathered.

The source area

The source area of sediment is the locality where rocks have been eroded to provide the sediment. Sediments are continually being produced by the processes of weathering and erosion and subsequently transported by water (fluvial processes), wind (aeolian processes) and ice (glacial processes).

The rock type exposed in a source area is particularly important as it determines the character of the resulting sediment in the modern and ancient depositional environment. The composition of the sediment can indicate the source area rock type, even if the source area has been completely eroded away. A conglomerate may contain pebbles of granite, limestone and sandstone, and these rock types were obviously available in the source area. An arkose (feldspar-rich sandstone) containing feldspar, quartz and mica may have come from a granitic source area. Quartz-rich sandstone containing well-rounded grains, on the other hand, probably represents the erosion of pre-existing sandstones. As quartz grains tend to be resistent to rounding, so well-rounded quartz grains are likely to have undergone many cycles of erosion, transportation and deposition, probably over tens of millions of years.

Sedimentary rocks can also be used to determine the distance from source area. Sediments tend to get thinner away from the source, and the sediment grains usually become smaller and more rounded. Alluvial fan sediments are a good example of how they become finer-grained away from the source area, and as a sedimentary environment, they are equally found close to the source of sediment supply, usually within a few kilometres (Fig. 4.2).

Environment of deposition

Figure 4.2 shows the common environments in which sediments are deposited, which are subdivided into three

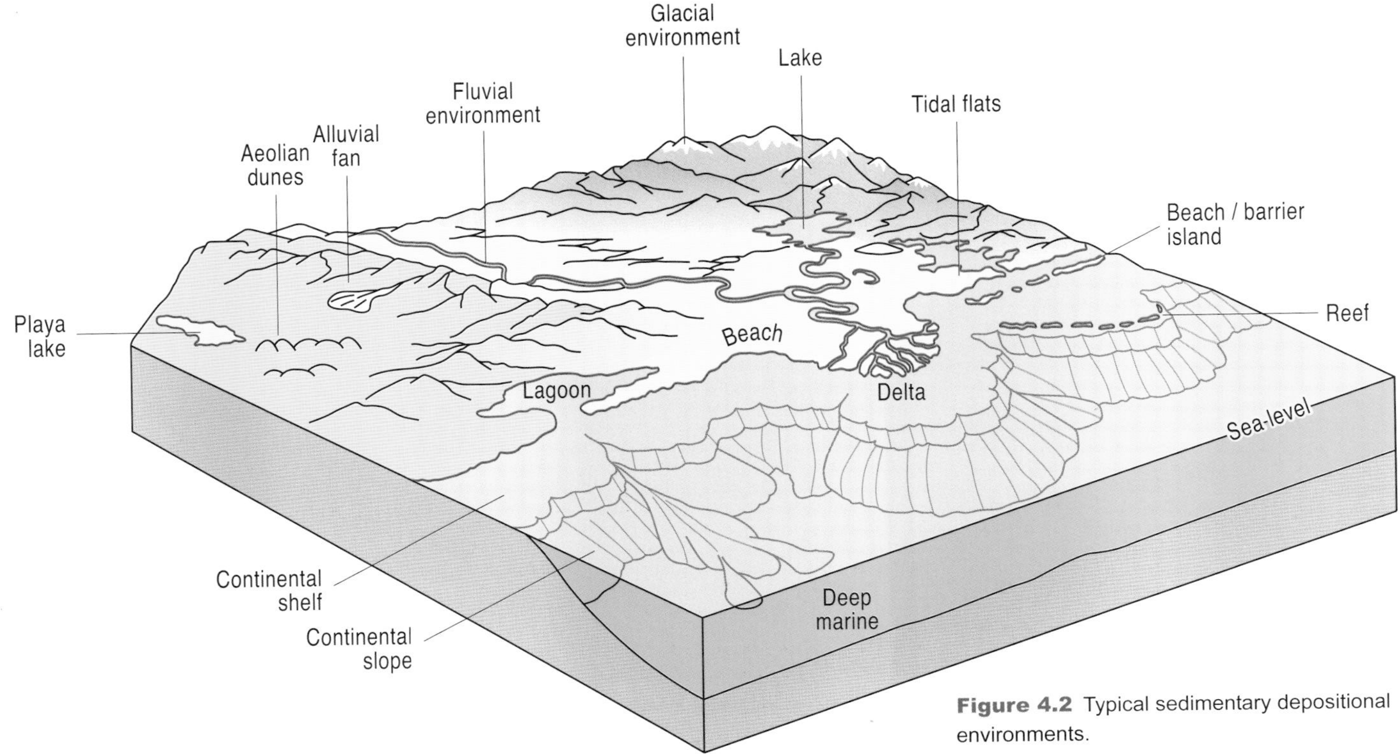

Figure 4.2 Typical sedimentary depositional environments.

basic categories of depositional environment: continental, marginal marine and marine. The different depositional environments and associated sediments are described in taking a journey from the mountains to the sea.

Continental environments

Sediments that occur in continental environments are found far enough away from the shoreline that they are not influenced by ocean tides or wave action. There are five commonly considered continental sedimentary environments: glacial, alluvial fan, river, lake, and hot deserts (Figs 4.2 and 4.3). All of these environments are closely linked, and climate plays an important role in determining water flux, sediment supply rates and glacial events. The interconnectedness between each of the continental subenvironments is linked to past and present climate. This inheritance requires the sedimentologist to carefully identify the consequences of climate change in the sedimentary rocks.

Glacial environments

At present, 10% of the land surface is covered with glacial ice, including glaciers, ice caps and the ice sheets of Greenland and Antarctica. In the

Figure 4.3 A Landsat image of central Iran, demonstrating the close interaction between many sedimentary environments, especially in non-marine settings.

geological past this has been significantly greater, with times even in the Precambrian when the Earth's surface was nearly entirely frozen over.

As a glacier moves down a valley in the mountains, it will carry all of the sediment that falls on its surface from adjacent cliffs or that is plucked from the underlying rocks. The rocks and sediment carried by a glacier will slowly smooth and polish the underlying rock. Where the ice contains large boulders, these can carve long scratches into the underlying rock, known as glacial striations. These help the sedimentologist to determine the direction of the glacier movement (Fig. 4.4).

At the end of a glacier, where the ice starts to melt it will drop its sedimentary load, known as glacial till. Till is poorly sorted and undifferentiated material ranging from clay size to boulders.

Alluvial fans

As streams emerge from mountains, at the mountain front onto flatter plains, they deposit broad, fan- or cone-shaped sedimentary deposits (Fig. 4.5A). As a stream's gradient decreases, it drops coarse-grained material. This reduces the capacity of the channel and forces it to change direction and gradually build up a slightly mounded or shallow conical fan shape (Fig. 4.5A, B). Typically alluvial fans in semi-arid climatic settings have a radius of 2–15 km. The sediments associated with alluvial fans are usually conglomerates, breccias, sandstones and lesser amounts of mudstone. They tend to be poorly sorted with the coarsest sediments found closest to the mountain front. The recognition of alluvial fan sediment in the geological record is important, as

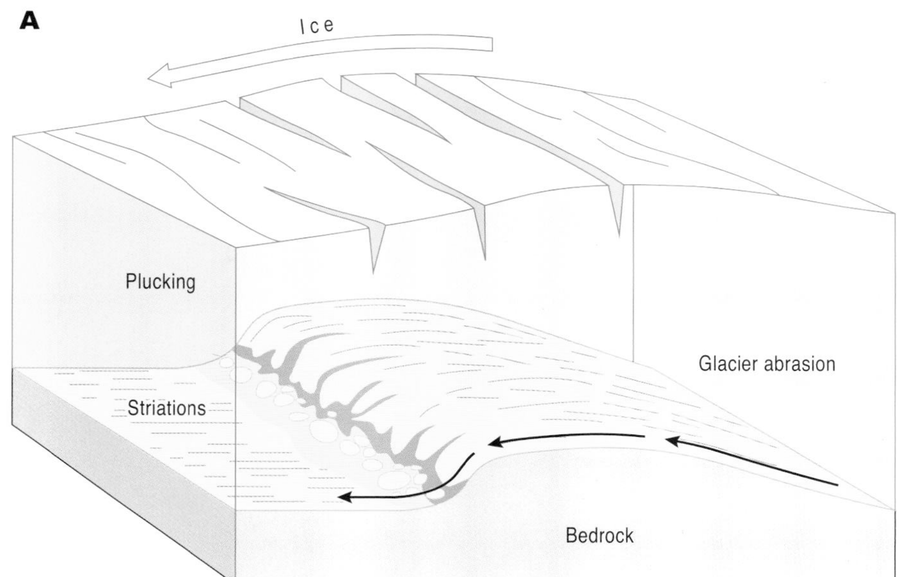

Figure 4.4 **A** Glaciers erode the bedrock through the processes of plucking and abrasion. These processes can slowly polish the bedrock surface, but equally when large boulders become trapped in the ice flow they can gouge grooves into the bedrock that are parallel to the direction of movement of the ice flow. **B** Large boulders incorporated into an ice flow of the Athabasca Glacier, which forms part of the Columbian Icefield, Canadian Rockies.

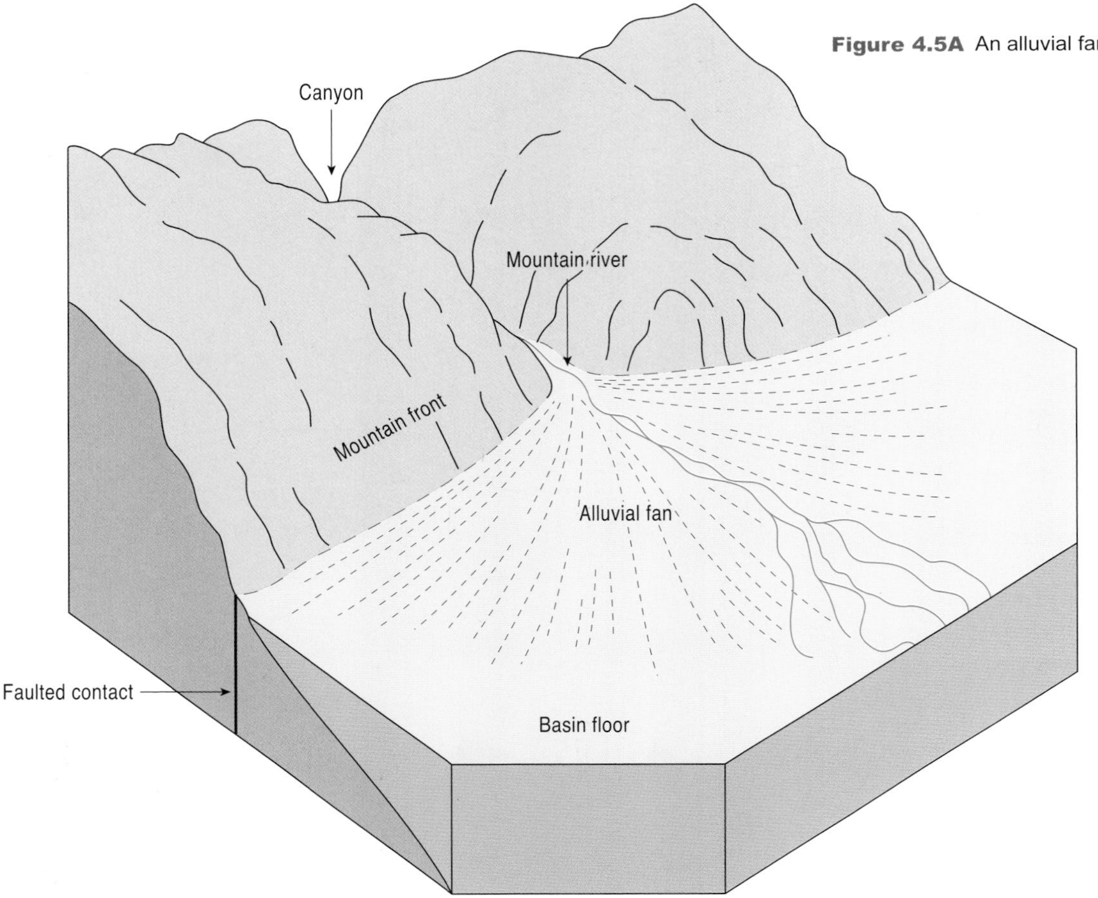

Canyon

Mountain river

Mountain front

Alluvial fan

Faulted contact

Basin floor

Figure 4.5A An alluvial fan forming at a mountain front.

Figure 4.5B An alluvial fan emerging from the Panamint Range in Death Valley National Park, California, USA. The toe of the alluvial fan has been truncated by a recent stream, creating the prominent scarps visible at the base of the fan. Sediment making up the fan is all locally sourced.

Figure 4.5C Miocene age alluvial fan at Riglos, Spanish Pyrenees. Beds of conglomerates were deposited in the proximal parts of an alluvial fan adjacent to the once tectonically active External Sierras. The alluvial fan has been partly eroded to leave these pinnacles of conglomerate.

alluvial fans are associated with mountain fronts that may have long been eroded away, and faulted sedimentary basins where associated subsidence can yield high preservation potential of the fan sediments (Fig. 4.5C).

Deserts

Deserts are areas of intense aridity with rainfall usually less than 250 mm per year and vegetation covering no more than 15% of their surface. Perhaps the most widely recognized deserts with large 'sand seas' are those forming in the subtropical latitudes between 20° and 30° north and south of the Equator. However, deserts can also occur on the inland side of mountain ranges, where a range shadow can form such as in Central Asia and the southern portion of the Andes in South America.

Well-rounded grains with a frosted surface, produced by the frequent grain collisions during wind movement, typify desert sands. The sandstones tend to be very well sorted due to the abrasive action of the wind, and usually red in colour due to iron oxide coatings of individual grains (Fig. 4.6A). Fossils tend to be absent from most desert sandstones due to the hostile environment, except for perhaps footprints and localized vertebrate bones.

The characteristic sedimentary structure of desert sandstones is

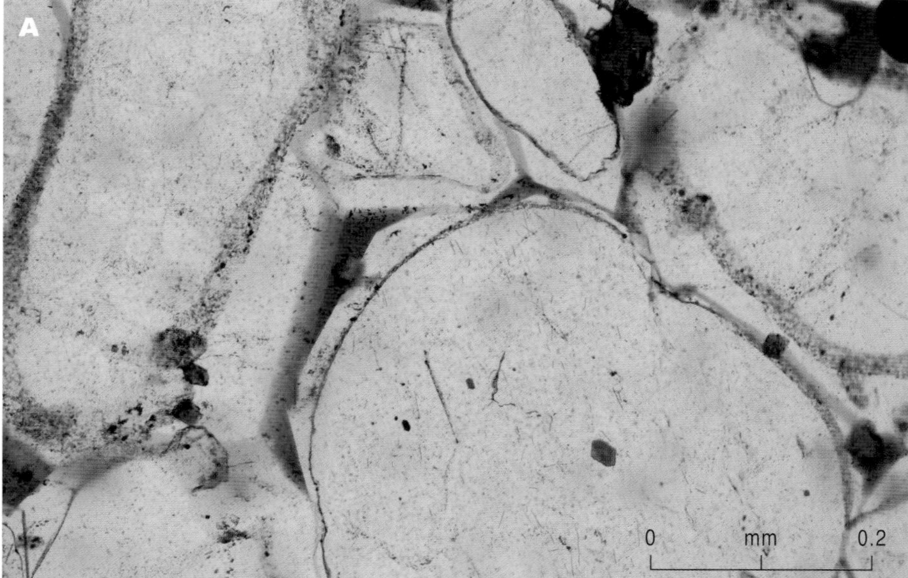

Figure 4.6 **A)** Photomicrograph of Triassic Penrith Sandstone, Appleby, Cumbria, UK. Rounded grains of quartz are indicative of aeolian sedimentary environments. The rounded grains have been cemented by a quartz cement that now forms an angular outline to the primary grains of quartz. The blue colour in this thin section is stained epoxy resin infilling pore space. **B)** Ventifacts are pebbles and cobbles that are abraded, pitted, etched, grooved, or polished by wind-driven sand. When ancient ventifacts are preserved without being moved or disturbed, they may serve as palaeo-wind indicators.

the large-scale cross-bedding and the absence of finer-grained sediments (see Fig. 3.14). Ancient desert dune sandstones are usually highly porous, so make excellent reservoirs for both water and hydrocarbons. Wind-blown sand grinds away at any pebbles, cobbles or boulders that may be exposed. Over time this sandblasting effect can produce faces or facets that may reveal ancient prevailing wind directions. Rocks whose surfaces have been faceted by the wind are known as wind-faceted pebbles or **ventifacts** (Fig. 4.6B).

Lakes

Lakes are highly diverse settings that vary greatly in size and may be fresh or saline, shallow or deep. In temperate climates, where water remains in the lake all year round, the lakes can be relatively quiet. Any sediment brought into the lake by rivers and streams will tend to settle out of suspension at the river outlet into the lake (Fig. 4.7A). Only fine-grained sediment makes it out into the centre of the lake, where the sediments

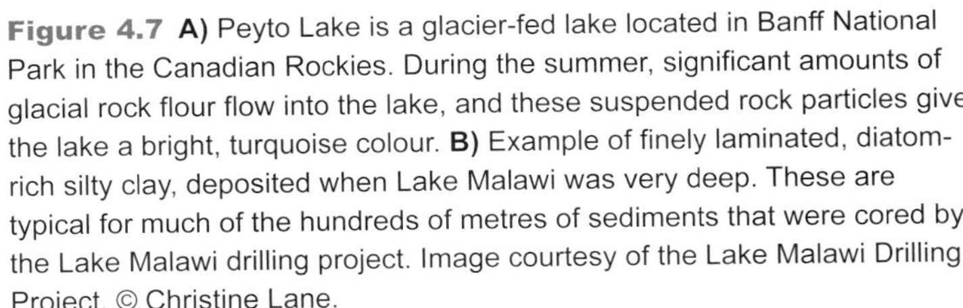

Figure 4.7 **A)** Peyto Lake is a glacier-fed lake located in Banff National Park in the Canadian Rockies. During the summer, significant amounts of glacial rock flour flow into the lake, and these suspended rock particles give the lake a bright, turquoise colour. **B)** Example of finely laminated, diatom-rich silty clay, deposited when Lake Malawi was very deep. These are typical for much of the hundreds of metres of sediments that were cored by the Lake Malawi drilling project. Image courtesy of the Lake Malawi Drilling Project. © Christine Lane.

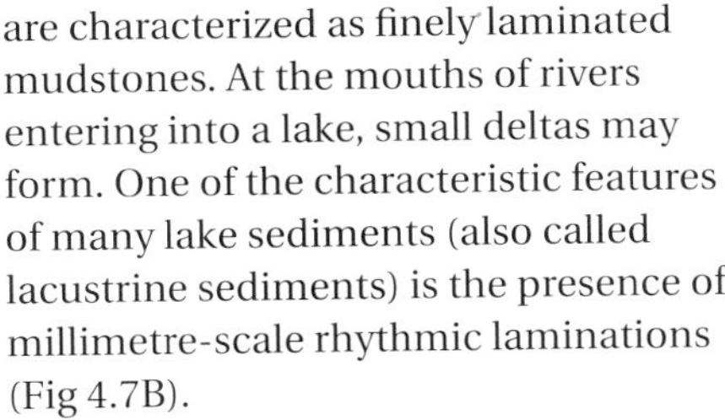

are characterized as finely laminated mudstones. At the mouths of rivers entering into a lake, small deltas may form. One of the characteristic features of many lake sediments (also called lacustrine sediments) is the presence of millimetre-scale rhythmic laminations (Fig 4.7B).

Although most lakes are freshwater, saline perennial and ephemeral lakes are common at low latitudes. Ephemeral lakes are common in semi-arid climatic settings and are often characterized by evaporitic sediments, mud cracks (Fig. 3.17) and fossil fish.

Rivers and flood plains

Modern and ancient river environments are a complex system of erosion, sediment transportation and deposition. Rivers are important in controlling the supply of sediment and water to other sedimentary environments, and their sedimentary record can tell the sedimentologist about the geological evolution of an area, including changes in sediment supply, variations in slope, and climatic changes within the catchment and drainage portions of a river.

River sediments tend to range from coarse conglomerates, through sandstones to mudstones. In general, sandstones associated with ancient rivers are usually sharp-based and cross-bedded, with some cross lamination and ripples (Fig. 4.8A). The sandstones can be lenticular in shape (the infills of river channels) and laterally pass sharply into finer-grained silts and mudstones associated with a floodplain. River sediments tend to lack fossils due to the erosive nature of the processes, but those associated are mostly plant material and skeletal remains of freshwater and terrestrial animals. When river sediments are

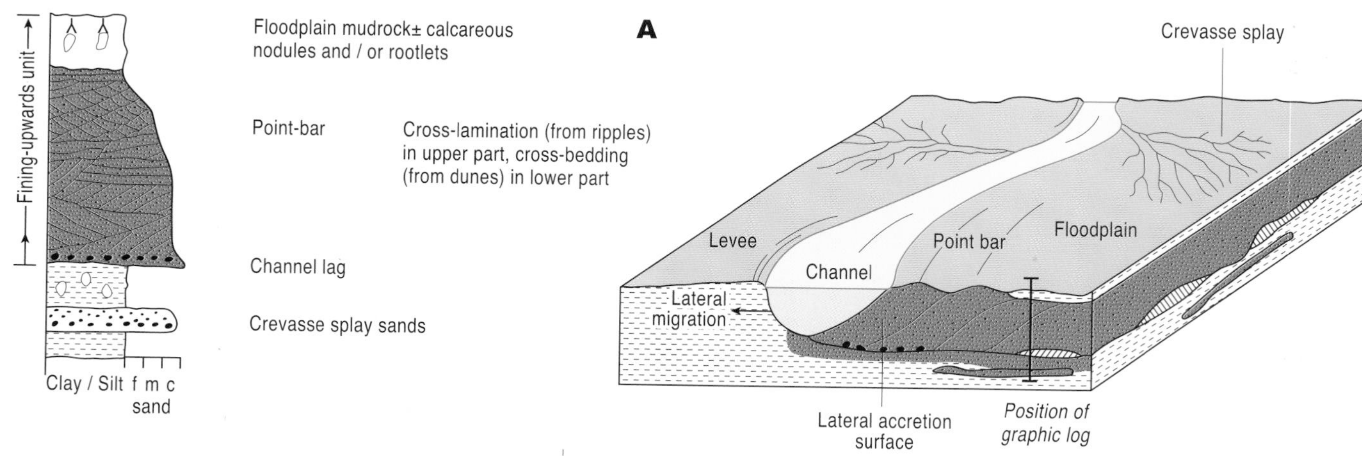

Figure 4.8 **A)** The subenvironments of a meandering river, and graphic log providing a vertical representation of the sedimentary units. **B)** Triassic age fluvial deposited sandstones and conglomerates of the Rio Gallo Gorge section, Aragon, Central Spain.

deposited in semi-arid climatic settings, iron commonly reacts with the oxygen in the surface or ground waters to produce rust-like iron oxide minerals (e.g. haematite), which give the sediments a reddish hue (e.g. Fig. 4.8B). Sedimentary strata that have this hue are informally known as 'red-beds'.

Marginal marine environments

Deltas

Where a river empties into the sea or a lake, the sediment may build a delta. Deltas are complex environments and are shaped by the interaction of waves, tides and river activity. Over time sediment is deposited at the coastline or lake margin, leading to deltas being subdivided into three main parts (Fig. 4.9A). The delta top or plain is where the river meets ocean and is dominated by

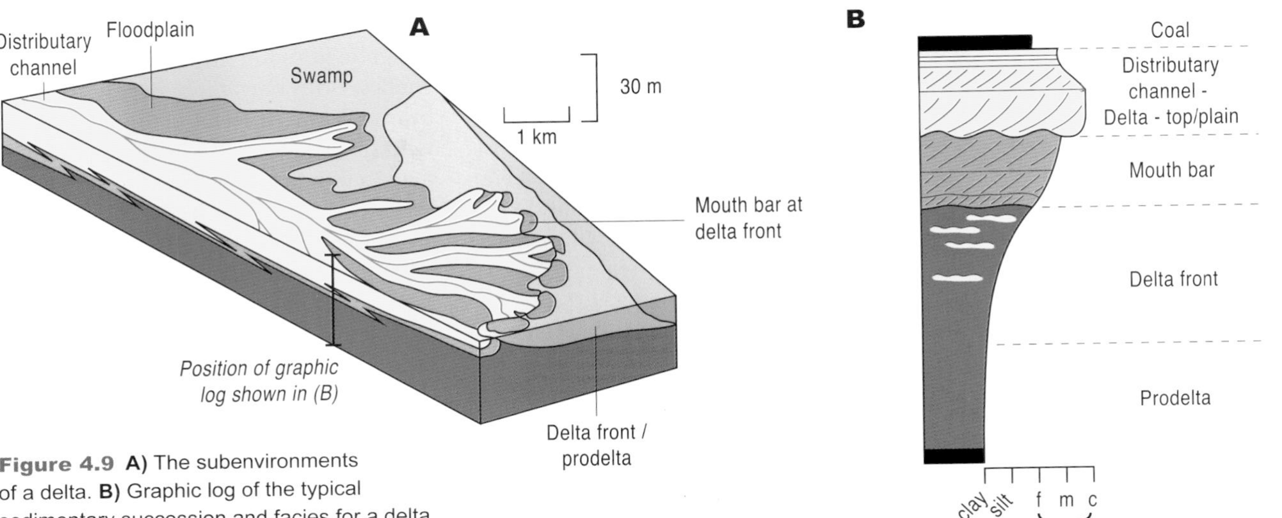

Figure 4.9 **A)** The subenvironments of a delta. **B)** Graphic log of the typical sedimentary succession and facies for a delta. **C)** Mississippi River delta with fine-grained sediment plume at the delta front supplying sediment to the pro-delta. The sediment plume from the Mississippi River is at the right, and the plume from the Atchafalaya River can be seen at the left. This is a true-colour image, acquired from the Moderate Resolution Imaging Spectroradiometer (MODIS) aboard NASA's Terra satellite on 5 March 2001. Image courtesy of NASA.

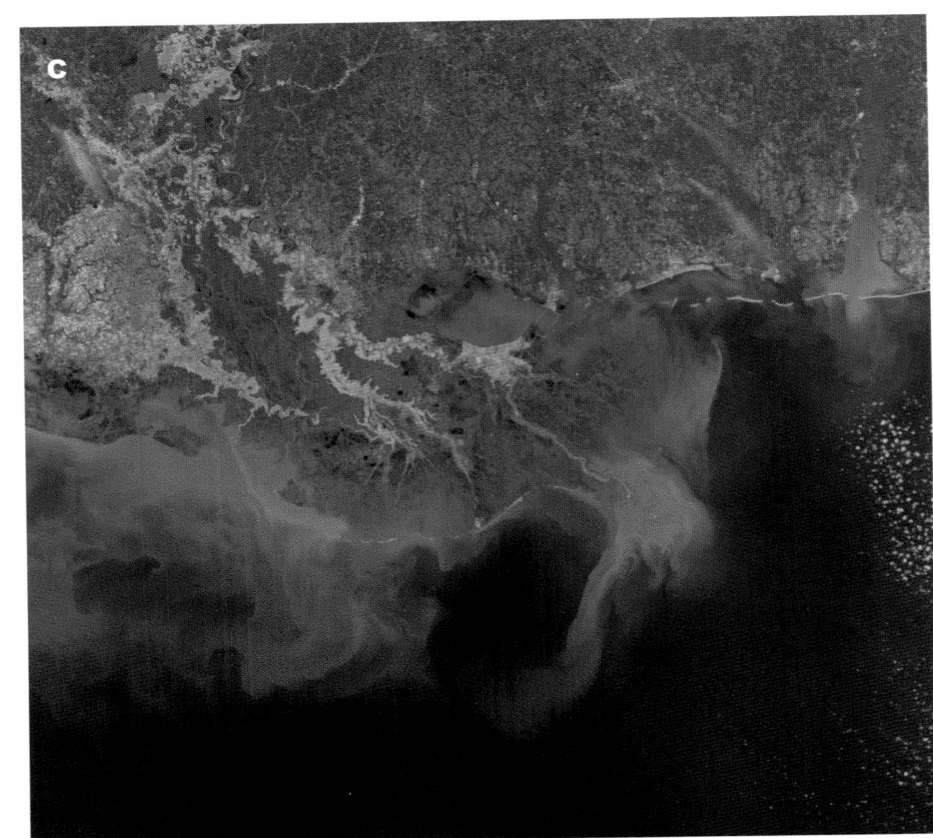

a variety of subenvironments including distributary channels, floodplains, swamps and lakes. Sediment types tend to be variable, from sands and sandstones and mudstones to coals with abundant plant material preserved (Fig. 4.9B). The delta front is the region where sediment carried by the distributary channels is deposited. Most delta fronts are dominated by sands with cross-bedding, ripples and some bioturbation. Sediment carried in suspension (clay) is carried farther offshore, often as a sediment plume, to be deposited on the prodelta (e.g. Fig. 4.9C). The prodelta is typically characterized by organic-rich, laminated and bioturbated mudstones, grading into deep-water local basin floor sediments.

The delta front is a region of rapid deposition and results in seaward migration of the delta. The sands of the mouth bar and delta front migrate over the finer-grained sediments of the prodelta. This produces thick (20–40 m) coarsening upward units (Fig 4.9B). The distributary channels of the delta top also migrate seaward and cut into the mouth bar sands.

Beaches

Moving along the coastline from a delta, wave action can transport sand long distances along the shoreline (Figs 4.2, 4.10). The continuous wave action tends to winnow out mud and silt to produce well-sorted sandstones with rounded grains. Ripples and cross-bedding may be preserved in ancient beach sands, but due to the high-energy environment only broken shelly material is likely to be preserved as fossils in such sediments.

Lagoons

A lagoon is a semi-enclosed quiet, shallow body of water between a barrier island or reef and the mainland (Fig. 4.2). Muds usually dominate lagoonal sediments, with occasional cross-bedded sands derived from storm wash-over of the barrier island. Lagoonal lime muds and limestones are also common and can be associated with small coral patch reefs that commonly grow in lagoons. Many modern hypersaline lagoons are home

Figure 4.10 A modern beach, Wharariki Beach, Golden Bay, New Zealand.

Figure 4.11 Modern stromatolites in Shark Bay, Western Australia. Stromatolites provide the most ancient records of life on Earth by fossil remains dating from more than 3.5 billion years ago.

to stromatolites, where extreme conditions due to high saline levels exclude animal grazing of the microbial mats and cyanobacteria, which trap sediment to form the mounds (Fig. 4.11).

Marine environments

Continental shelves

Moving into slightly deeper waters away from the shoreline, the continental shelves are covered in sediment that represents the staging post between erosion of the continents and final deposition in the deep ocean (Fig. 4.2). Sediments on the continental shelves are dispersed by a complicated mix of tides, waves and ocean currents where water depths range from 5 to 200 m. Widespread sands, silts and muds can be deposited preserving a variety of sedimentary structures. Hummocky cross-stratification and erosion surfaces may occur in many sands, produced from the passage of storm waves. Planar cross-bedding and ripples, often with mud drapes, indicating tidal conditions, are common in shallower parts of the shelves. In ancient shelfal sediments marine fossils can be abundant, as well as diverse types of trace fossils.

Often found in close association with sands and muds of the continental shelf are carbonate sediments. As seawater is oversaturated in carbonate and warm waters favour precipitation, the subtropical and tropical shallow continental shelves are dominated by carbonate-rich sediments. In carbonate environments, the nature of sediment depends upon the water depth and the type of shelly fauna that lives in the warm waters (Fig. 4.12).

A distinctive carbonate sedimentary rock found is oolite (Fig. 2.7). Oolites are usually formed in shallow, highly agitated intertidal environments. The mechanism of formation starts with a small fragment of sediment acting as a 'seed', e.g. a piece of a shell. Strong intertidal currents wash the 'seeds' around on the seabed, where they accumulate layers of chemically precipitated calcite (carbonate) from the supersaturated seawater. They typically form in agitated waters where the ooids are moved around as large dunes and ripples by wave action and tidal and storm currents.

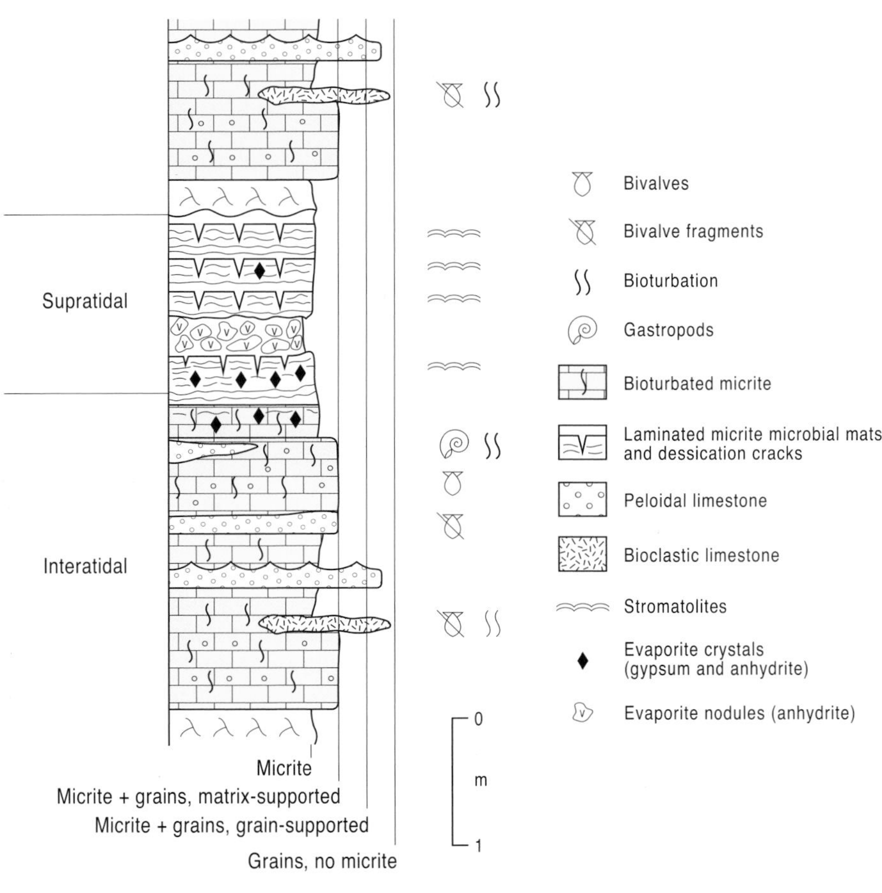

Figure 4.12 Graphic log of a typical intertidal to supratidal carbonate ramp succession.

Reefs

Coral reefs are one of the most familiar of modern carbonate environments, and widely reported upon as a key indicator for ongoing global climate change (Fig. 4.13A). The term 'reef' is usually applied to a carbonate buildup, predominantly comprising coral but also encrusting coralline algae. The organisms that have made up reefs have changed through the geological past, with cyanobacteria and microbes particularly important for constructing stromatolite bioherms in the Precambrian and Cambrian, while, for example, sponges dominate in the Triassic and Jurassic.

The majority of reefs occur along the shelf margin in wave-agitated waters (Fig. 4.13B). However, smaller patch reefs can occur in open lagoons, or atolls developed in the open ocean on submerged volcanic islands (Fig. 4.13C). The growth of modern coral reefs is controlled by many factors such as water depth, salinity, water temperature (around 25°C), wave action and absence of silt and clay. Such controls of coral growth help with understanding ancient sedimentary environments and the palaeoclimatic setting. The structure of modern shelf margin reefs with a reef front/slope, reef-crest and back reef provides a very distinctive subdivision of sediment types. The reef front is a steep slope, often with a talus slope of coarse reef debris. The reef-crest is the main site of coral growth, passing shoreward into quiet waters with lagoonal carbonate muds and possible patch reefs. A similar pattern of sedimentary facies can be seen in many ancient coral reefs, such as the classic Permian El Capitan reef from the Guadalupe Mountains, Texas, USA (Fig. 4.13D).

Figure 4.13A Iolanda reef in Ras Muhammad nature park, Sinai, Egypt. Photo by Mikhail Rogov.

a) Fringing reef

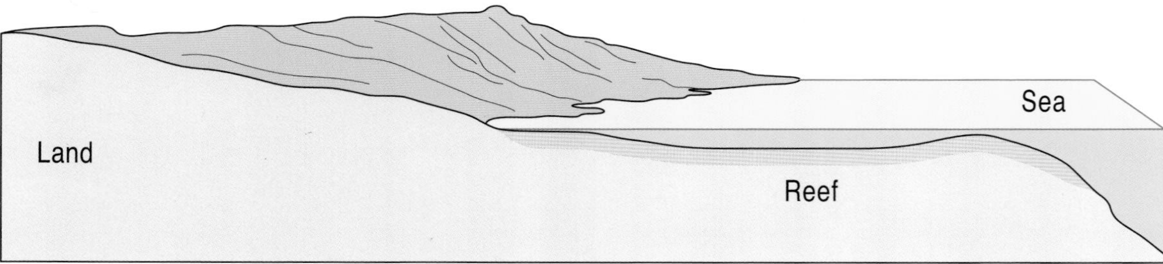

b) Atoll

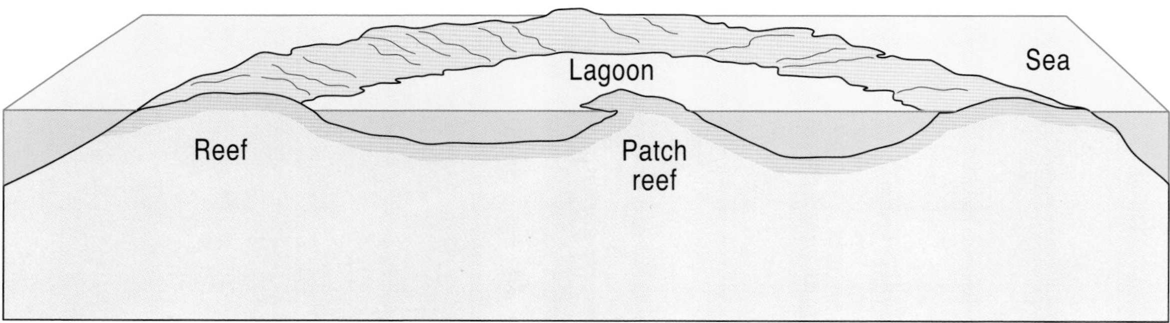

c) Barrier reef

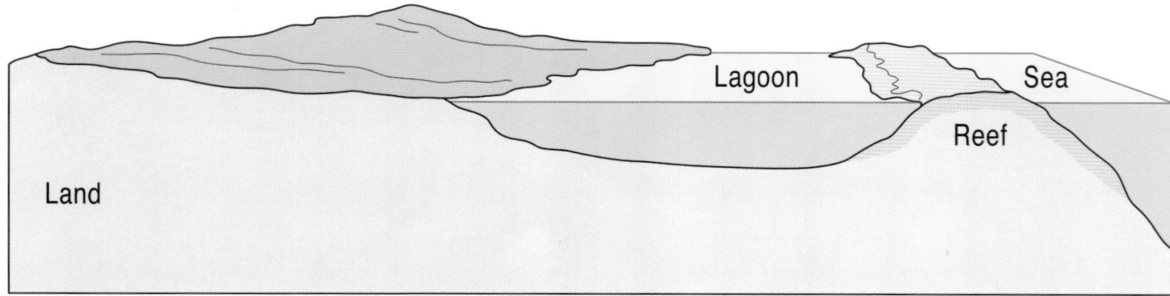

Figure 4.13B Main types of coral reef.

Figure 4.13C Satellite image of Atafu, a coral atoll surrounding a central lagoon in the south Pacific Ocean, 500 km north of Samoa. Image courtesy of the Image Science & Analysis Laboratory, NASA Johnson Space Centre.

Figure 4.13D The El Capitan ancient reef. El Capitan is the southern part of the Guadalupe escarpment, an ancient Permian limestone reef that forms the present-day Guadalupe Mountains, Texas, USA.

Deep sea

Venturing into even deeper waters, and to some of the deepest parts of the world oceans at depths greater than about 500 m, there are many kinds of deep-sea sediments. The main source of sediment is from the continental shelf, where transportation by slumps and slides can lead to turbidity currents. Further offshore pelagic limestones, cherts and mudstones primarily dominate the environment.

The occurrence of slides and slumping at the continental shelf slope is frequently induced by earthquakes, but rapid sedimentation, storm-wave loading and sea-level changes can also produce instability (Fig. 4.14). The generation of sediment gravity flows by slumps and slides is one of the most important ways of transporting sediment, namely silt and sand, tens to hundreds of kilometres into the deep marine environment. Turbidity

Figure 4.14 Slump-related folds preserved in Miocene age sediments from the Tabernas Basin, southern Spain. High sedimentation rates and localized earthquakes can destabilize the sediments to cause slumping.

currents are the most important of the sediment gravity flows, where sediment is maintained in suspension by fluid turbulence. As a turbidity current starts to decelerate due to a change in slope or horizontal flow expansion, a definite succession of internal structures and grain size changes occurs. The sedimentologist Arnold Bouma (1932–2011) first identified a sequence for dividing deep-water turbidites into intervals; this is known as the **Bouma Sequence**. The Bouma sequence is divided into five distinct layers labelled A through E, with A being at the bottom and E being at the top (Fig. 4.15). Each layer described by Bouma has a specific set of sedimentary structures and a specific lithology, with the layers overall becoming finer-grained from bottom to top. Most turbidites found in nature have incomplete Bouma sequences because their formation does depend on grain size and rate of deposition (see Figs 3.1, 3.4 and 3.5).

Frequently associated with turbidites in deepwater environments are pelagic sediments resulting from the settling of microscopic, calcareous or siliceous shells of phytoplankton and zooplankton in the open ocean (Unit E of Bouma sequence; Fig. 4.2). Modern pelagic carbonates are

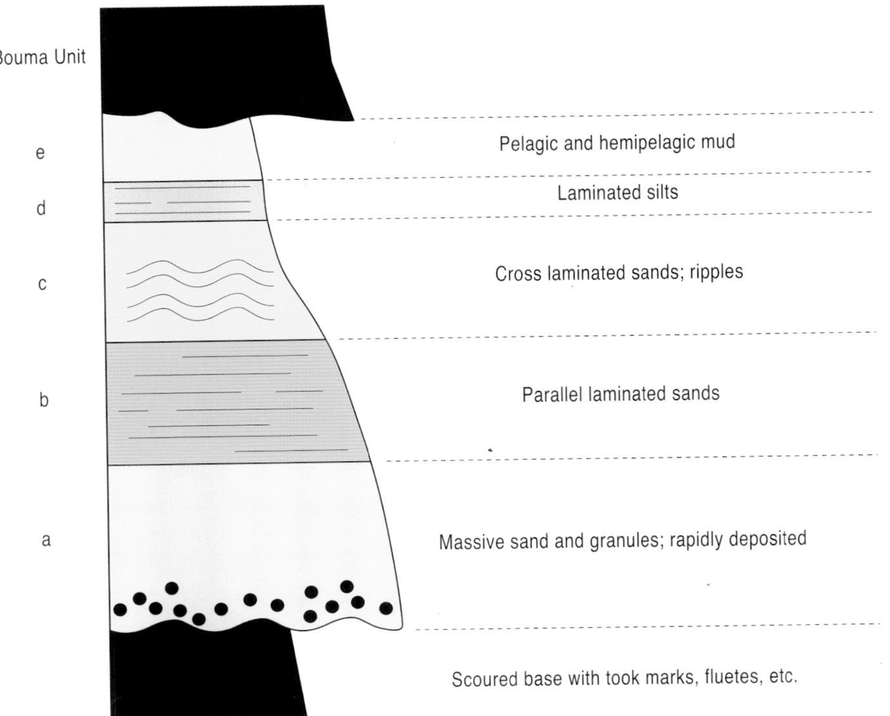

Figure 4.15 An idealized Bouma Sequence (after Arnold Bouma, 1932–2011).The Bouma sequence specifically describes the ideal vertical succession of structures deposited by low-density (i.e. low sand concentration, fine- to medium-grained) turbidity currents. Synthesized from original source. Bouma, Arnold H. (1962) *Sedimentology of some Flysch deposits: A graphic approach to facies interpretation.* Elsevier, p. 168.

predominantly made up of coccoliths and foraminifers. The Cretaceous chalks of northwest Europe are composed largely of coccoliths, and are thus pelagic limestones (see Fig. 5.4).

Extra-terrestrial sedimentary environments

The dynamic and ever-changing nature of the sedimentary environments on Earth contrasts markedly with those on other planets. However, sediment is also known on the surface of many other planets and their satellites, notably Mars, Venus, and Titan, Saturn's largest moon. The remarkably sophisticated imagery from the many NASA's Mars Exploration Rover lander missions has helped to transform our understanding of the planet and raised many questions about the sedimentary environments and physical processes on Mars and other planets.

Perhaps the most intriguing aspect of the sedimentary record of Mars is the evidence for a wet ancient climate, despite the unlikely occurrence of the current Mars climate supporting any quantity of liquid water at the surface. Extensive channels, distributary networks and palaeolakes once existed on the surface of Mars. The layered outcrop known as 'Shaler' identified by the NASA Mars rover Curiosity mission in 2012 has been interpreted as being deposited in a fluvio-lacustrine setting with readily identifiable cross-bedding that may have been created in a fluvial environment when water existed at the surface on Mars (Fig. 4.16).

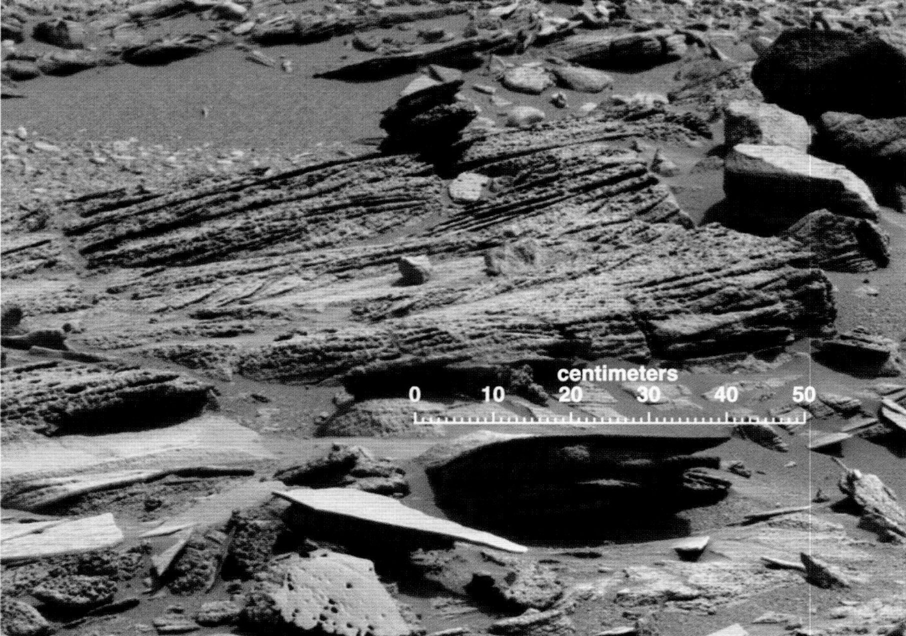

Figure 4.16 An image from the Mast Camera (Mastcam) on NASA's Mars rover Curiosity. Cross-bedding in an outcrop called 'Shaler' is clearly visible. The image has been white-balanced to show what the rock would look like if it were on Earth. Also see in Figures 3.10 and 3.11 how the cross-bedding may have been created. Image credit: NASA/JPL-Caltech/MSSS.

The modern climate on Mars is dominated by aeolian processes, and these types of sediment are by far the most widespread of all the sedimentary strata. Indeed, any flat surface seems to support ripples, dunes of reddish dust and black sand and numerous ventifacts (see Fig. 4.6B). Both the presence and morphology of sand dunes are sensitive to subtle shifts in wind circulation patterns and wind strengths, and are thought to be influenced by changes in Martian orbital parameters. The spatial distribution of aeolian sand relates to patterns of sedimentary deposition and erosion of source materials, giving clues to the sedimentary history of the surrounding terrain (Fig. 4.17). Furthermore, aeolian transport of dust is what gives Mars's sky most of its colour, and is the reason why Mars is often called the red planet.

Figure 4.17 This colourful image of windblown sediment on Mars is from the Noctis Labyrinthus region. The series of pale ridges record the occurrence of transverse aeolian ridges. These differ from a second set of more active dunes that are darker in colour and often record smaller sets of ripples cross-cutting the dunes. The dark dunes are made up of grains composed of iron-rich minerals derived from volcanic rocks on Mars, unlike the pale, quartz-rich dunes typical of Earth. Image Credit: NASA/JPL/University of Arizona.

5 Fossils and sediments

The importance of fossils in sediments for interpretation of past depositional environments has been mentioned briefly in chapter 3. However, they provide fundamental information about the evolution of life on Earth and the basis for correlation of sedimentary strata.

The occurrence of shells, bones and plant fossils in sedimentary rocks is clear evidence of the fact that the nature of organisms living on earth has changed through time. Many fossils found resemble animals and plants living today, whereas others capture our imagination with prehistoric life, for example with dinosaurs that roamed the earth for 165 million years, and new discoveries are regularly followed with public enthusiasm.

Our understanding of fossils is based on the accumulated knowledge from previous generations of palaeontologists and sedimentologists, who carefully recorded the identity and distribution of fossils from numerous geological exposures or samples from wells, and recorded their findings in the scientific literature. When the regional or global distribution of fossils through geological time is taken into consideration, we can gain important insights into plate tectonics, community migration, and palaeoclimatology.

Types of fossils found in sediments

Body fossils

Typical fossils, such as shells, bones, claws and teeth are called 'body' fossils because they represent the actual remains of once living organisms. These are the most commonly recognized of all fossil types. Most body fossils are the remains of a once living creature, e.g. ammonites, crinoids, bivalves, corals, dinosaur bones (Fig. 5.1). This is not always the case, and trilobites

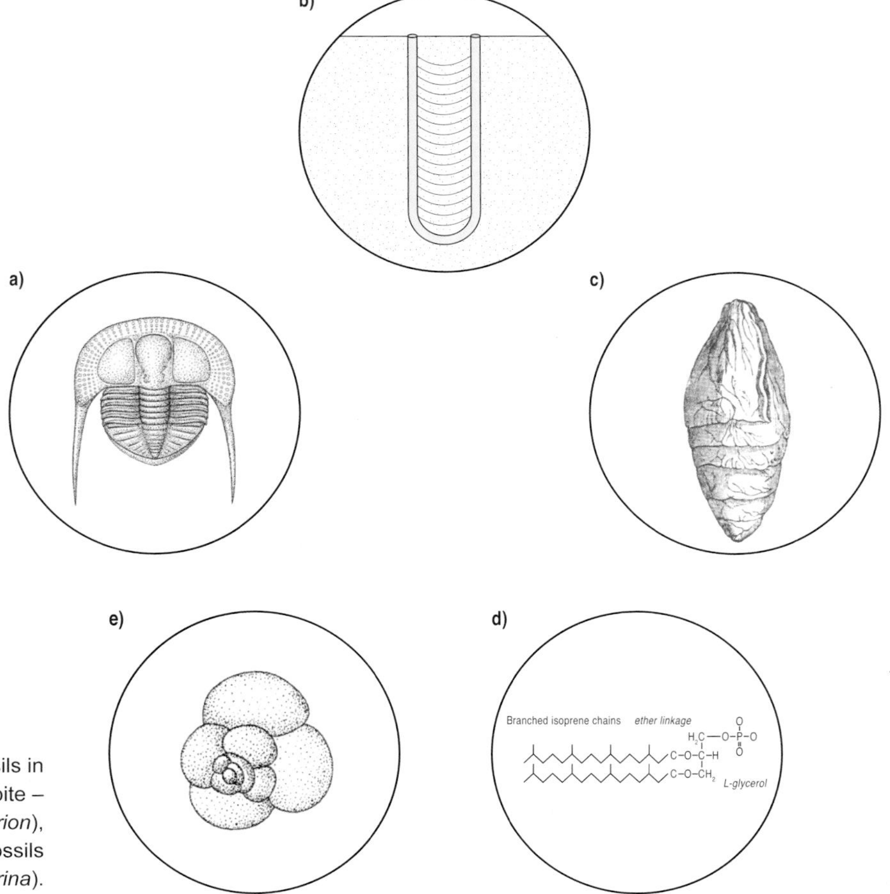

Figure 5.1 Types of commonly found fossils in sedimentary rocks. **A)** Body fossil (e.g. a trilobite – *Trinucleus*); **B)** Trace fossil burrow (*Diplocraterion*), **C)** Coprolite; **D)** Chemical fossil; **E)** Microsfossils (e.g. a foraminifera – *Globigerina*).

are an example that shed their exoskeleton as they grew; many trilobite fossils are the preserved moults that are abundant throughout the Palaeozoic. Body fossils can be used in **biostratigraphy**, but no single group fulfils all the criteria necessary for correlation and age dating of sediments. Some fossils, such as ammonites, are particularly useful in correlation of Jurassic age sediments, whereas trilobites are used for correlation in the Cambrian. Many body fossils can greatly assist the sedimentologist with palaeoenvironmental interpretations (Fig. 5.2).

Trace fossils.
The study of trace fossils, the preserved impressions of biological activity in sediments, is exceptionally important, as described in chapter 3. Trace fossils provide direct evidence for fossil behaviour preserved in the sediments and are excellent indicators of past sedimentary environments and associated environmental conditions (e.g. oxygen levels, nutrient abundances, rate of sedimentation).

Coprolites
A **coprolite** is fossilized faeces (Fig. 5.1). They can be classified as a trace fossil but are quite rare because they tend to decay rapidly. Typically nodular or contorted in appearance, coprolites are composed of pulverized indigestible remnants of an organism's food, such as portions of scales, bones, teeth, shells, plant material and pollen or spores. They yield important information about the dietary habits, feeding behaviour and coexistence with other organisms. It was Mary Anning (1799–1847), a

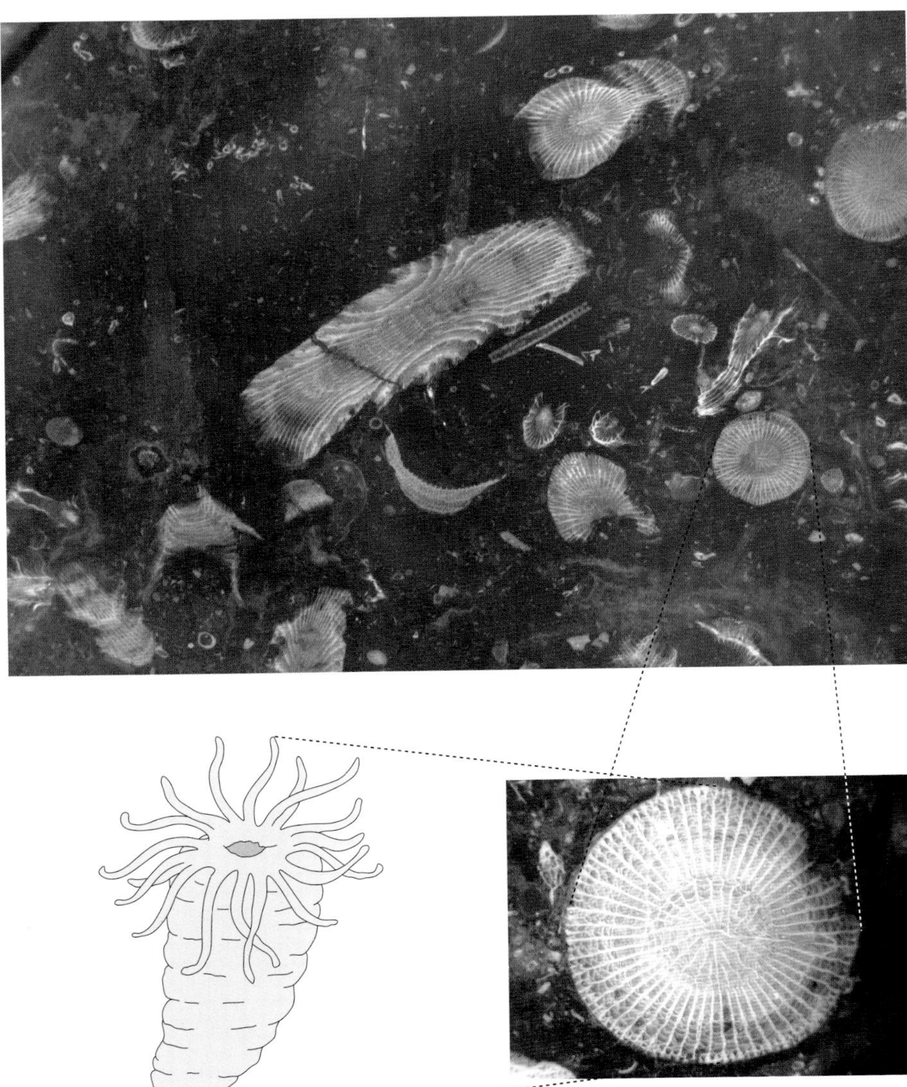

Figure 5.2 Extinct solitary rugose coral of *Dibunophyllum bipartitum* that is found in the Carboniferous Frosterley Marble of North Pennines, UK. These creatures had curved horn-shaped skeletons and tentacles that filtered out particles of organic matter from the seawater. When they died, their skeletons collected on the sea floor and were covered by limy ooze (calcium carbonate), which hardened into dark grey limestone. The fossils provide evidence that during the Carboniferous Period northern England basked in a tropical climate and was periodically covered by warm, shallow seas. Coral redrawn after sketch by Elizabeth Pickett ©North Pennines AONB Partnership.

palaeontologist, who made important discoveries in the Jurassic stratigraphy of the Lyme Regis area, Dorset. Her discoveries included the first ichthyosaur (a large marine reptile) skeleton to be correctly identified, and her keen observations played a key role in the discovery that coprolites, known as bezoar stones at the time, were fossilized faeces (Fig. 5.3).

Chemical fossils

Chemical fossils are chemicals found in rocks that provide an organic signature for ancient life (Fig. 5.1). Nucleic acids (e.g. DNA) proteins and carbohydrates do not survive for long in sedimentary environments and through geological time. The majority of chemical tracers come from chlorophyll and related organic molecules. Fossil fuels found in sediments, such as crude oil, coal and natural gas, are the result of biological activity and contain chemical fossils. Coal deposits represent plant material that once grew, with crude oil and natural gas formed primarily from algae and zooplankton (see chapter 6). Many Archean fossils (4000–2500 million years ago) are only preserved as chemical traces in the rocks and are thought to represent simple non-nucleated single-celled organisms called **Prokaryota**.

Microfossils

Microfossils are fossil remains that are too small to be seen with the naked eye or a hand lens (usually no larger than 1 mm). They are examined using a microscope (Fig. 5.1). Many sediments contain microfossils and they are important biostratigraphic, palaeoenvironmental and palaeoceanographic indicators. Microfossils are particularly important in biostratigraphy due to their abundance, durability and global distribution. Microfossils can be divided into four areas of study:

- calcareous microfossils, e.g. foraminifera and coccoliths;
- siliceous microfossils, e.g. diatoms and radiolarians;
- phosphatic microfossils, e.g. conodonts; and
- organic microfossils, e.g. pollen and spores.

The Late Cretaceous chalks of NW Europe are made from predominantly calcareous microfossils called coccolithophores, which rained down on the sea floor from the sunlit waters above. Each minuscule individual has a spherical skeleton called a coccosphere, formed from a number of interlocking calcareous discs called coccoliths. After death, most coccospheres and coccoliths collapse into their constituent parts and accumulate to form chalk, a type of fine-grained limestone (Fig. 5.4).

Figure 5.3 Ichthyosaur coprolites from the Lower Jurassic, Whitby, North Yorkshire. © Mike Marshall, *Yorkshire Coast Fossils*.

Using fossils to tell the time: biostratigraphy

At the time of the British industrial revolution in the late eighteenth and early nineteenth centuries, factories demanded ever more coal for their steam engines. It was decided to build a network of canals to transport coal, and accordingly engineer William Smith (1769–1839) was hired to survey the excavations. The canals provided

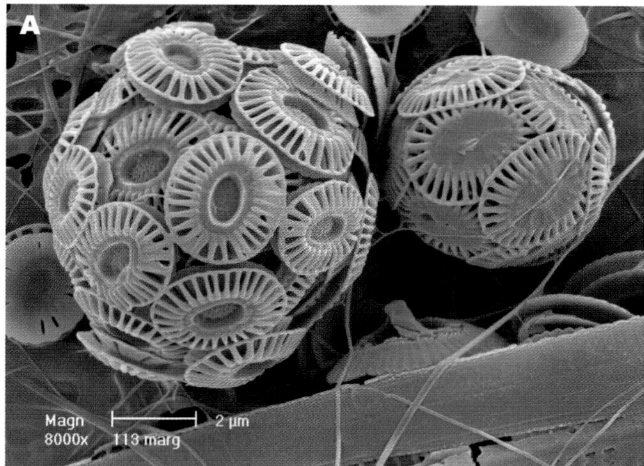

Figure 5.4 **A)** coccolithophore morphology and; **B)** Chalk cliffs at Flamborough Head, North Yorkshire, UK.

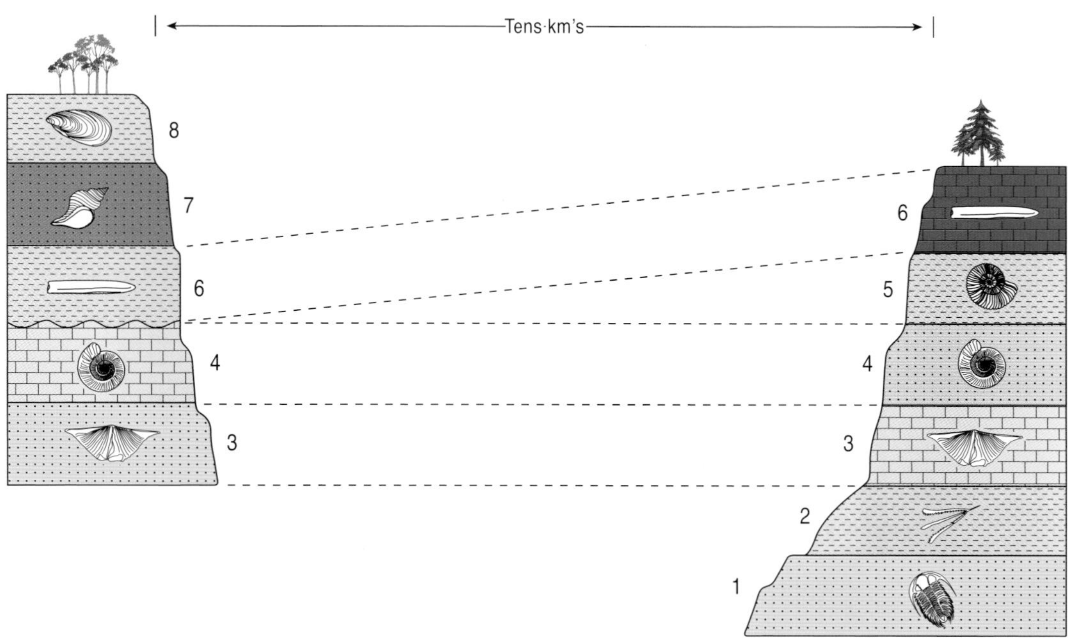

Figure 5.5 Principle of biostratigraphy. The fossils are useful, as sediments of the same age can look completely different because of local variations in the sedimentary environment. For example, one section might have been made up of clays and marls while another has more chalky limestones, but if the fossil species recorded are similar, the two sediments are likely to have been laid down at the same time.

fresh exposures of geology, previously covered by vegetation. Smith learned to recognize the different sedimentary strata and identify the fossils that they contained. He realized that they were arranged in a predictable pattern and that the various strata could always be found in the same relative positions. Additionally, each particular stratum could be identified by the fossils it contained, and the same succession of fossil groups from older to younger rocks could be found in many parts of England. Smith's observations have been repeated by stratigraphers at millions of locations around the world and gave rise to the **principle of fossil succession** (Fig. 5.5).

Biostratigraphy and **geochronology** underpin much of geoscience. No matter what aspect of geology one is working on, the most common question posed by geologists is 'what age is it'? Biostratigraphy and geochronology provide the framework for answering that question. Biostratigraphy is the study of the temporal and spatial distribution of fossil organisms. The limited stratigraphic range of many fossil taxa is used for correlation, typically by means of biozonation schemes (i.e. intervals characterized by a species or group of species; Fig. 5.6). When interpolated with numerical age information derived from radiometric dating, biozones and the divisions of the geological timescale containing biozones (periods, epochs, stages, etc.) can be given chronological values.

A practical application of biostratrigraphy is for oil and gas exploration. When wells are drilled through underground rock formations, microfossils in particular are used in correlation and to help with dating for locating and exploitation of hydrocarbon reserves.

Exceptional preservation

Fossil sites with exceptional preservation, sometimes including preserved soft tissues, are known as **fossil Lagerstätten**. Frequently soft and delicate tissues are retained, and these can provide information about the morphology of the organisms; in many cases these fossil Lagerstätten are the only examples in the fossil record of soft-bodied taxa. In order to preserve such soft tissues the rate of decay must be arrested or stopped altogether at the time of death.

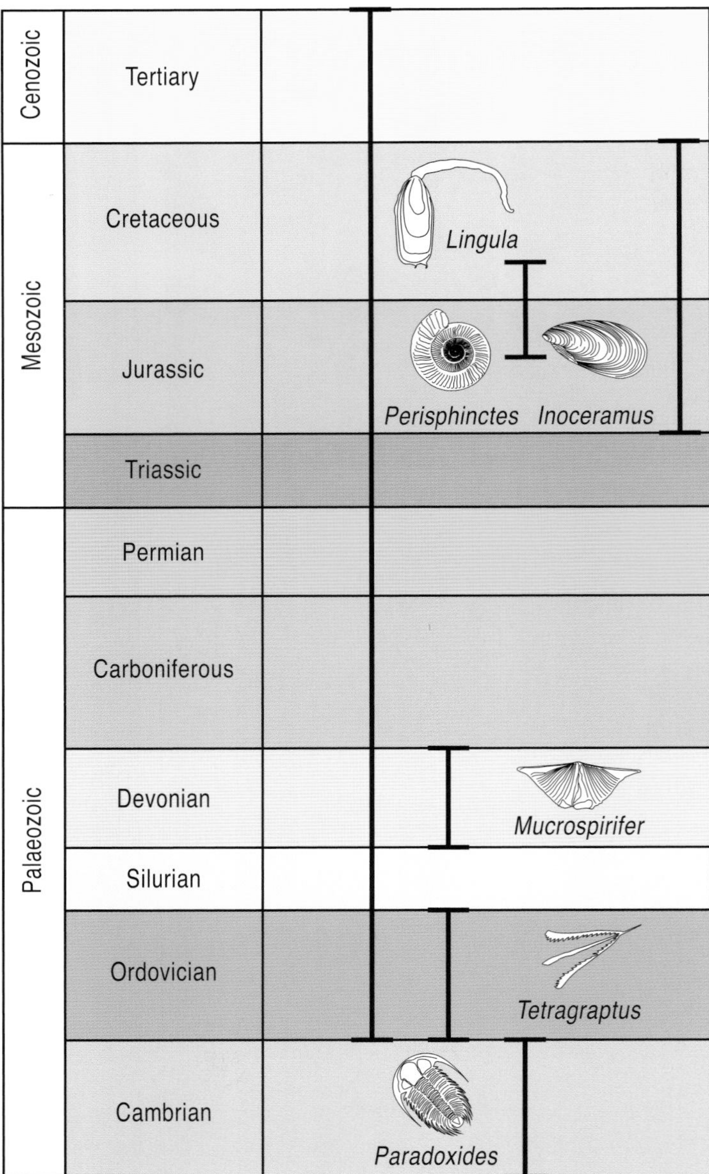

Figure 5.6 The geological age ranges of several marine invertebrates. The trilobite *Paradoxides*, the graptolite T*etragraptus*, the brachiopod *Mucrospirifer*, the ammonite *Perisphinctes* and the bivalve *Inoceramus* all make excellent index fossils for correlation and biozonation. However, in comparison the brachiopod *Lingula* is easily identified and widespread but has a very long geological range from the Ordovician to Recent and is of little use for biostratigraphy.

Lagerstätten span geological time from the Cambrian to the present day and store critical data on the evolution of life on our planet. Perhaps some of the most interesting and significant are the Cambrian Lagerstätten, which provide an insight into a critical point in the evolution of life on Earth, the Cambrian Explosion of animal body plans. The most famous is the Middle Cambrian Burgess Shale in the Canadian Rocky Mountains. Here over 125 fossil genera have been discovered, dominated by arthropods. The Burgess Shale fauna is thought to have once lived in a marine shallow-water shelfal setting, but the animals were swept into a deeper-water abyssal setting by large-scale slumping and turbidity currents. The fossils are found in shales, and largely preserved articulated with soft tissues. This degree of preservation was only possible due to the rapid burial in an anoxic environment with minimal bacteria, thus slowing decomposition.

Equally famous, but for differing reasons, is the younger Jurassic Solnhofen Limestone of Bavaria, Germany. The Solnhofen Limestone is very fine-grained and thinly bedded, which made it particularly useful for making lithographic printing plates. The extensive quarrying of the region in the nineteenth century yielded many important fossils, but the most important is that of *Archaeopteryx lithographica* (Fig. 5.7). *Archaeopteryx* is a genus of early bird that represents the transition between dinosaurs and modern birds. The key to exceptional preservation in the Solnhofen Limestone is the fine-grained nature of the sediment, and that the limestone

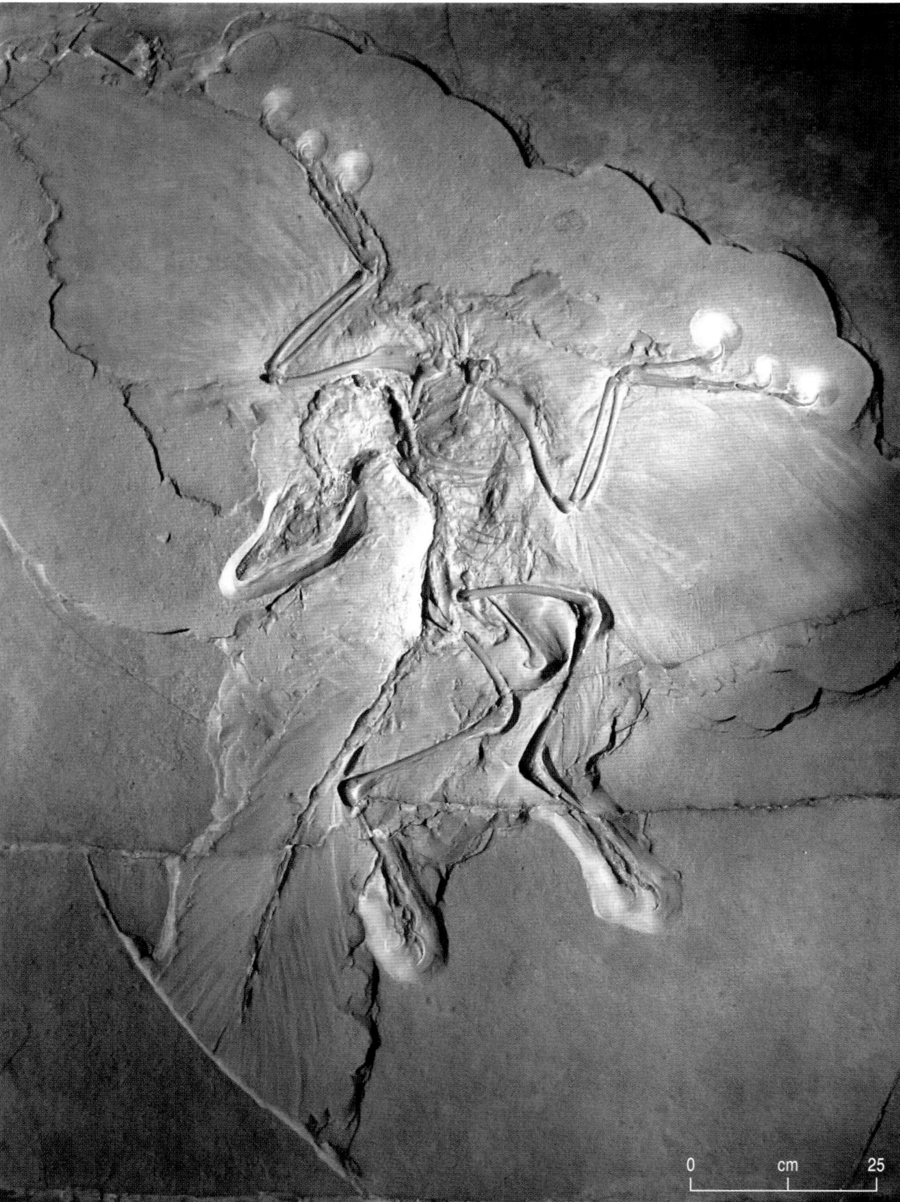

Figure 5.7 *Archaeopteryx lithographica* fossil, around 147 million years old. The fine-grained nature of the limestone and high salinity of the stagnant lagoonal waters allowed exquisite preservation of the fossil, including impressions of wing and tail feathers. Specimen displayed at the Museum für Naturkunde in Berlin, which is a real specimen and not a cast. © H. Raab. Creative Commons Attribution – from Wikipedia.

was deposited in stagnant lagoonal waters into which animals fell or were washed during storms. The high salinity of the waters aided preservation.

What fossils do not tell the sedimentologist

Fossils preserved in sedimentary strata demonstrate bias according to their morphology and the environment in which they once lived. This inherent difficulty affects the way we interpret past sedimentary environments (Fig.5.8). The key, of course, is to understand and recognize these biases and account for them and how they affect things, and to make allowances as far as possible for their effects. To become a fossil, the remains of an organism must not decay away to nothing, but instead be buried in sediment (e.g. sand, mud). This must lie undisturbed for long enough and under enough pressure from new sedimentary strata for the material to become mineralized and turn into a fossil. This immediately reveals some of the biases and the need for caution when interpreting fossils in sedimentary rocks. We commonly see a bias towards fossils with hard body parts (e.g. shells and bones); those once living in marine settings and in particular low-energy environments; and organisms that were abundant and widespread are also more readily preserved as fossils.

Organisms that have hard body parts such as a shell, bones and teeth are more readily fossilized than those with soft body parts. For example, sedentary animals such as bivalves and corals have more robust shells,

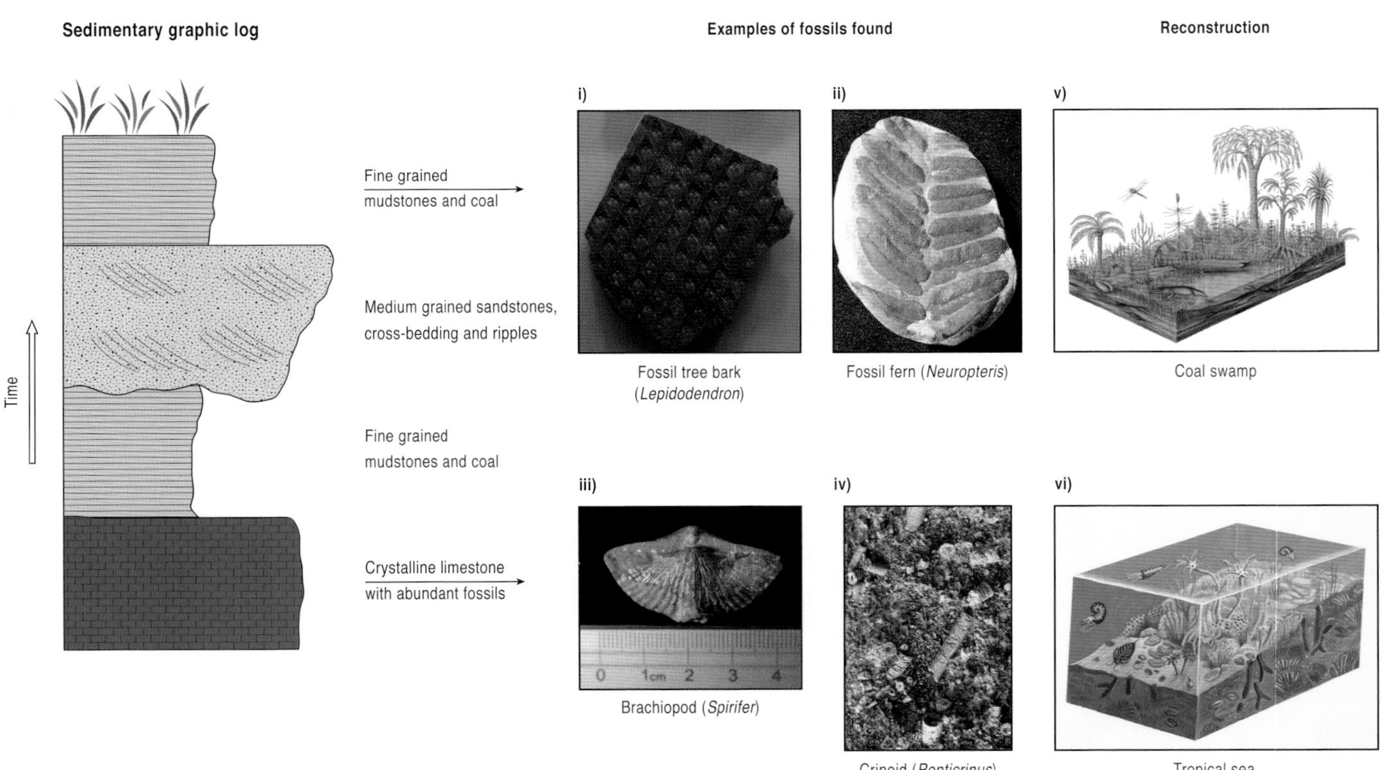

Figure 5.8 A sequence of sedimentary rocks records the changing environments through time. The combined understanding of the sedimentary rocks and fossils found allows for reconstructions of the ancient sedimentary environment to be created. The stratigraphy, fossils and reconstructions are for the North Pennines, UK, during the Carboniferous.Carboniferous coal swamp © Elizabeth Pickett; tropical sea is by Elizabeth Pickett © BGS, NERC.

increasing their chances of preservation. More active predators such as fish and reptiles have a lower preservation potential due to their less robust skeletons, but also being mobile, could more readily escape being buried by sediment. The preservation of animals with only soft tissues happens only under exceptional circumstances, so these fossils are rare.

Organisms that once lived in a marine setting have a better preservation potential than those living on land. This reflects the increased erosion and lower rates of deposition in continental settings, with less opportunity for burial and preservation. The rates of sedimentation in different sedimentary environments are a further factor in the preservation and distribution of fossils. Slow sedimentation rates commonly result in poor preservation because the remains of organisms are left to decay on the sea floor with biogenic degradation and scavenging by other creatures. However, organisms may be preserved where slow sedimentation rates are combined with anoxic conditions.

Knowing the limitations of interpreting fossils found in sedimentary rocks allows the sedimentologist to better understand ancient sedimentary environments, and in doing so we can make sense of why certain fossils are found more commonly in particular sedimentary rocks (Fig.5.8).

DETERMINING THE ENVIRONMENT

The kinds of animals and plants living in a particular area will depend on the local environment. By looking at the geological and biological information recorded in a sedimentary rock that preserves a fossil, you can determine the type of environment where a fossil organism once lived.

- **Sediment composition and grain size** tell you what sort of surface the animal once lived upon or within.
- **Sedimentary structures** such as ripples and cross-bedded strata indicate the organism lived in moving water where sediment was moved by currents. The occurrence of mud cracks and fine-grained sediments could indicate periodic drying out and a harsher environment for fossil preservation.
- **Broken shells** may indicate pounding by waves, as on a beach or created during storm conditions.
- **Diagenesis** including the rate of burial, the starting pore fluid chemistry and later mineral replacements within sediment can significantly influence the preservation potential of a fossil. Conversely, exceptional preservation can indicate unique conditions of the sediments and diagenetic processes.
- **Murky waters,** with low light levels and lots of suspended sediment particles, can be determined by the type of sedimentary rock and associated sedimentary structures. Fine-grained mudstones are made of tiny clay particles that can be easily suspended in water. Filter-feeding organisms such as corals are not usually found in muddy waters because the suspended sediments clog their filters.
- **Oxygen levels are critical to all organisms.** If there is insufficient oxygen in the water, relatively few organisms may have lived in such an environment. However, in oxygen-deficient waters organic material in the sediments will not readily decompose and organisms that have perhaps died elsewhere and floated into the environment can be well preserved.

6 The riches from sedimentary rocks

The use of sedimentary rocks

Since the Industrial Revolution of the 1800s society's appetite for energy and natural resources has increased almost unabated. Nearly every substance used in our daily lives has either been mined from Earth's crust or manufactured using energy from fossil fuels. Our modern lives have become dependent on a vast variety of natural resources, and many of these are associated with or found in sediments and sedimentary rocks.

The sedimentary natural resources can be grouped into four categories: i) Water resources; ii) energy resources; iii) metallic resources; and iv) nonmetallic resources commonly used in the construction industry. This chapter does not include all the different natural resources associated with sediments and sedimentary rocks but hopes to provide a flavour of the wide variety.

Groundwater

Frequently neglected as one of our most precious natural resources is water. It is what makes our planet unique and supports life on Earth. Many communities obtain water from rivers, lakes and reservoirs, with often aqueducts or canals to redirect surface waters. A more economic resource of water is that found in the subsurface, known as **groundwater**, which provides most of the water needs in human society. Less than 3% of the total global water on Earth is fresh water, about 75% is frozen in glaciers, approximately 25% is groundwater, and only about 0.005% is found in lakes and rivers. Groundwater is vulnerable to overuse and pollution. Frequently in cities next to large rivers, they may pump water from the ground because groundwater is commonly less polluted and more economical to extract.

Responding to gravity, water percolates down into the ground through cracks, pores and spaces found in soil, sediments and sedimentary rocks. The rate of groundwater flow tends to decrease with depth because sedimentary rock pore space becomes filled with cement and increasingly compacted due to the weight of the overlying rocks. Furthermore, sedimentary rocks have different amounts of pore space due to their depositional environments and diagenesis (Fig.6.1; Table 6.1).

Nearer the ground surface, water usually only partially fills the pore space, leaving some space filled with air. This region of the subsurface, in which water only partially fills the pore space available, is called the undersaturated zone or vadose zone (Fig. 6.2). Deeper down, water completely fills or saturates the pores, and this region is known as the saturated zone or phreatic zone. These two zones are separated by a horizon known at the **water table** (Fig.6.2).

Natural groundwater is commonly slightly acidic because of dissolved carbon dioxide (CO_2) from the atmosphere or from soil gases. This acidity of groundwater can dissolve limestone along joints and bedding planes, opening up caves or caverns. Most caves are thought to form at or near the water table, and as the water table drops, caves empty of water and become filled with air. Over time the caves can be joined through underground networks, but this usually only occurs where there are thick layers of limestone in the subsurface.

Energy resources

Coal fuelled the industrial revolution of the eighteenth and nineteenth centuries, but oil and gas have become the favoured fuel for the twentieth century. Often called 'black gold', oil plays a crucial part in global economics with just about every industry critically dependent upon it and countries willing to wage war over this black gold.

Oil and gas

Most oil forms from marine organic matter (mostly from planktonic organisms), but some can form from dead algae in lacustrine settings. As these fine-grained organic-rich mudrocks are buried they contain the raw materials from which oil and gas can

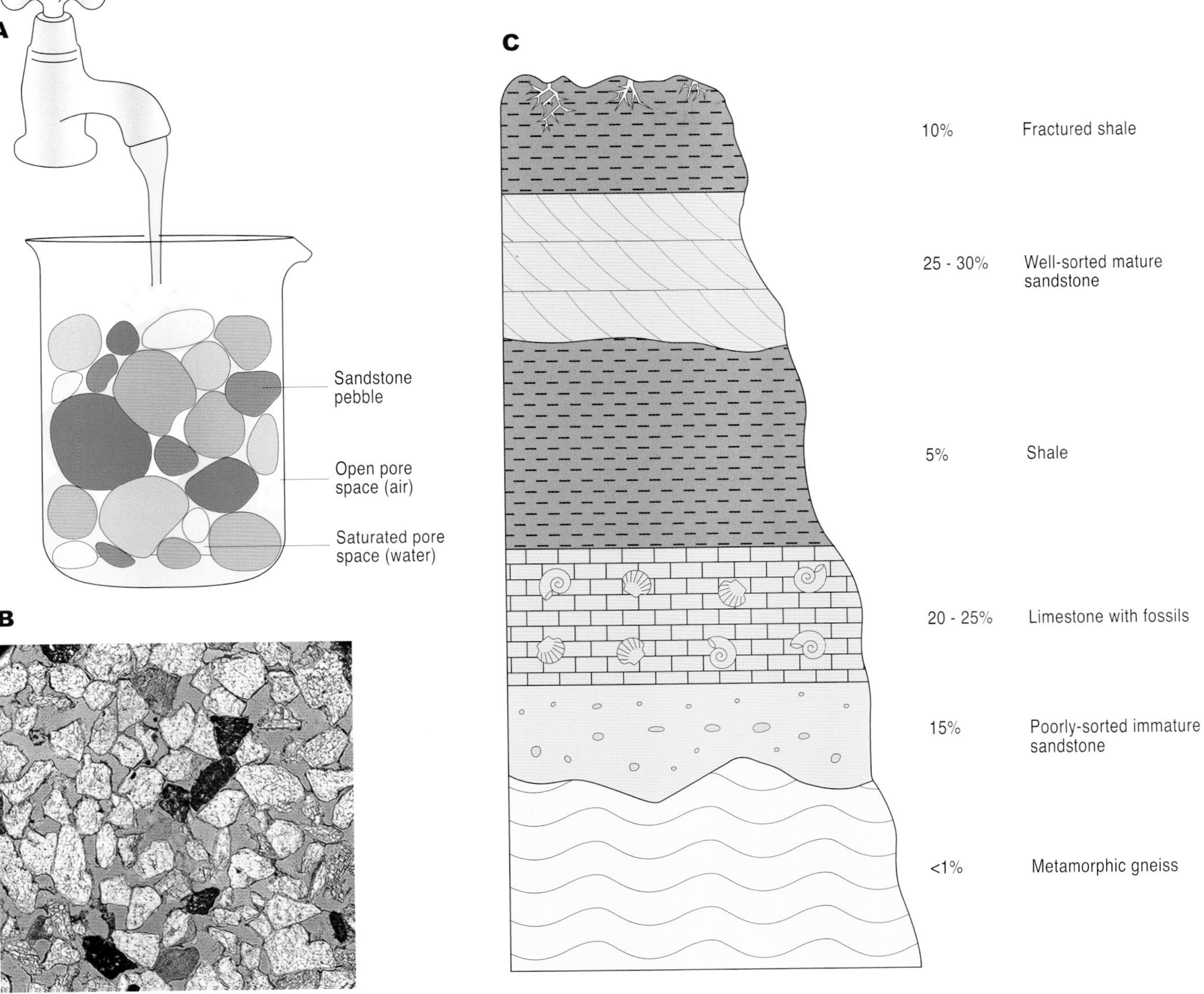

Figure 6.1 **A)** The pebbles in this gravel do not fit together well, which creates pore space. Pore space, more commonly called porosity (total percentage of pore space in a rock or sediment), allows water to be stored between the pebbles and/or grains. This is important for storage of water in aquifers and hydrocarbons in reservoirs. **B)** Photomicrograph of a porous sandstone, where the blue colour is a stained epoxy resin dye added to the rock to identify porosity. The sandstone thin section is from the Triassic Skagerrak Formation of the North Sea and was deposited by an ancient braided river. **C)** Different types of sedimentary rocks have very different porosity, and this can be due to sorting, presence of fossils and grain size.

Table 6.1 Typical porosity and permeability encountered in sediments and rocks.

Sediment type	Porosity (%)	Permeability
Gravel	30 - 40	Excellent
Sand (clean)	40 - 50	Good to excellent
Silt	35 - 50	Moderate
Clay	35 - 80	Poor
Glacial till	10 - 20	Poor to moderate
Rock type	**Porosity (%)**	**Permeability**
Conglomerate	10 - 30	Moderate to excellent
Sandstone		
Well-sorted, little cement	20 - 35	Good to very good
Average	10 - 20	Moderate to good
Poorly sorted, well-cemented	0 - 10	Poor to moderate
Shale	0 - 30	Very poor to poor
Limestone, dolomite	0 - 25	Poor to good
Fractured and cavernous limestone	Up to 50	Excellent
Crystalline rock		
Unfractured	0 - 5	Very poor
Fractured	5 - 10	Poor
Volcanic rock	0 - 50	Poor to excellent

be generated. As the organic-rich mudrocks undergo burial diagenesis the organic material slowly undergoes a transformation into waxy molecules called kerogen. A mudrock containing kerogen is often called oil shale. As the kerogen is subjected to deeper burial and increased pressure and temperature, it will start to release oil. Significant amounts of oil only begin to form at temperatures over 50°C and the largest quantity of oil is formed as the kerogen is heated to temperatures between 60 and 150°C. At still higher temperatures oil becomes thermally unstable and breaks down or 'cracks' to natural gas. Thus oil itself forms only in a relatively narrow range of temperatures, called the **oil window** (Fig. 6.3).

Once formed, over 90% of the oil and gas in sediments will escape at the surface through natural seepage (Fig. 6.4). Modern oil seeps do not concentrate enough oil to be economic. Instead, most sedimentologists look for oil and gas in **reservoir rocks**, rocks that contain an abundant amount of easily accessible oil and gas. To be a reservoir rock, the rock

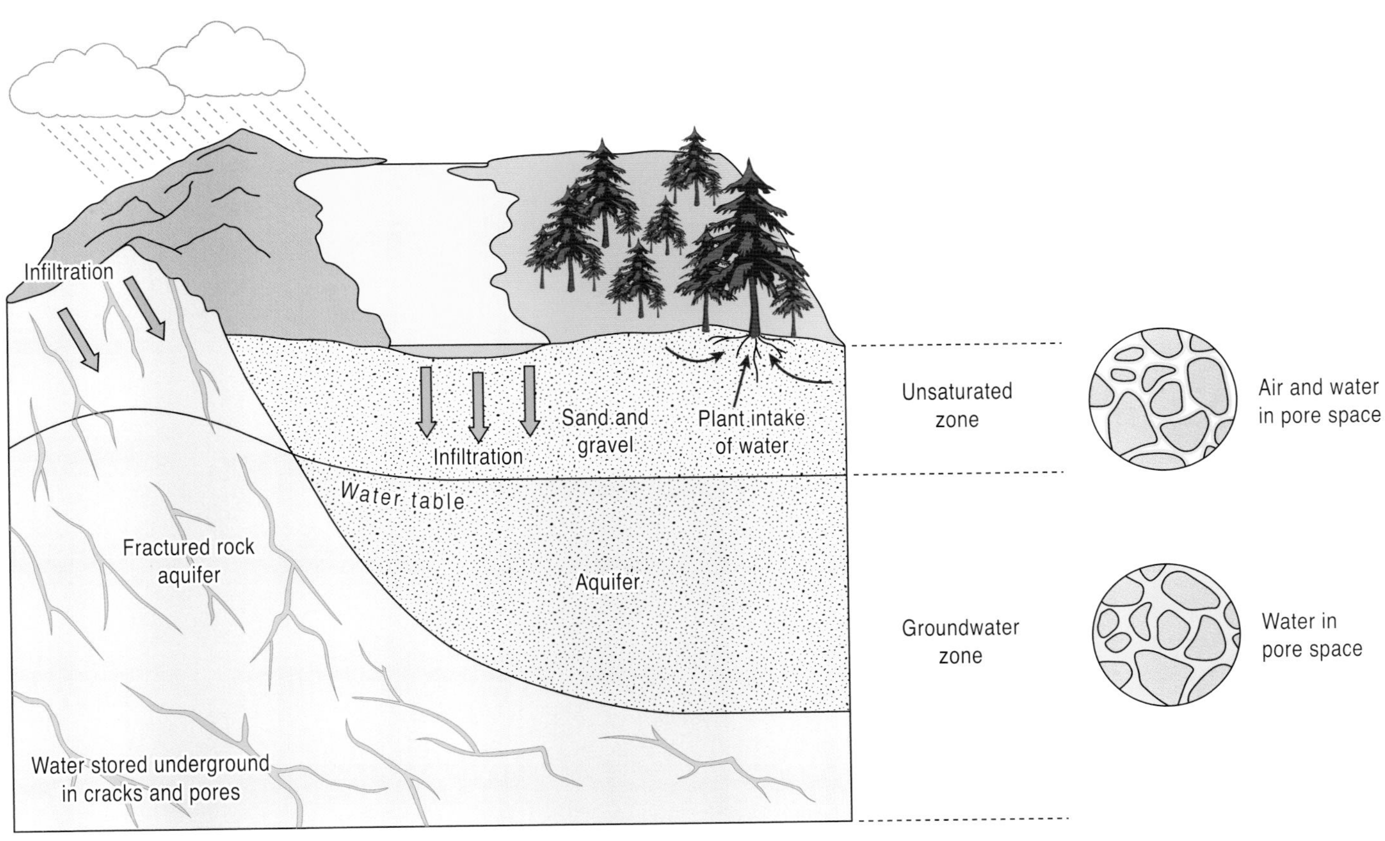

Figure 6.2 Water can enter permeable sediments and bedrock through cracks and connected pore space. The water table is defined as the top of groundwater in the subsurface. It separates the unsaturated zone from the groundwater or saturated zone beneath. http://www.gov.pe.ca/environment/index.php3?number=1015822&lang=E [Accessed 2014.]

must have sufficient pore space for the oil and/or gas to reside. Sandstone and limestone are the most common reservoirs, as they tend to be sufficiently porous and permeable to allow the oil and gas to migrate and become stored under the correct burial conditions.

To fill the pore spaces in the reservoir rock, the oil and gas must first migrate from the source rock into the reservoir rock, which commonly takes place over millions of years (Table 6.1). Oil and gas are less dense than water, so they tend to rise towards the Earth's surface (Fig. 6.5). However, for the oil and gas to accumulate in the reservoir to economic levels there must be a geological trap. There are two key components to a geological trap (Fig. 6.5). First, a seal, a relatively impermeable rock such as salt, mudstone or unfractured limestone must lie above the reservoir rock and stop the oil and gas from rising any further (Table 6.1). Second, the seal and reservoir must be arranged to allow the oil and gas to collect, and this usually involves folding of the sedimentary rock layers or faulted traps as the most common types.

To extract the oil and/or gas from the reservoir an oil well is drilled, that is simply a hole drilled into the ground to a depth where it penetrates the reservoir. Oil flows from the reservoir into the well and then

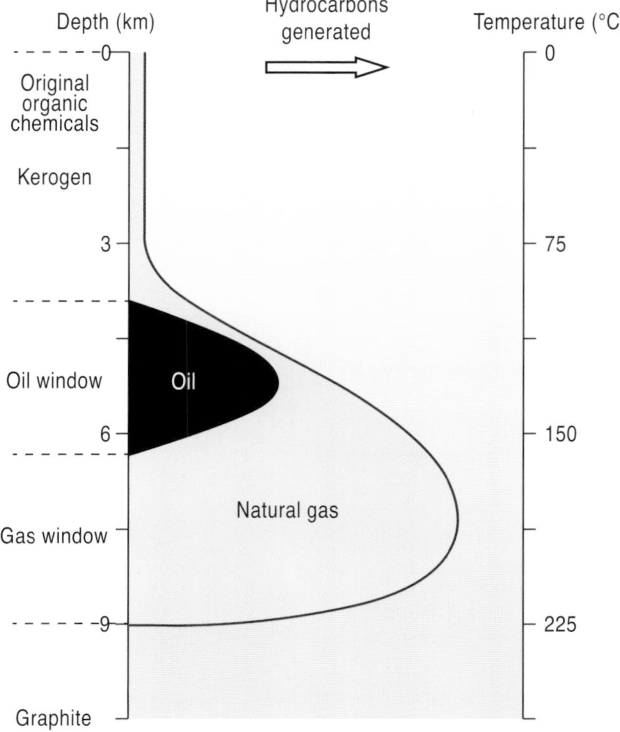

Figure 6.3 The 'oil and gas window'. As temperature rises with increasing burial depth sediments rich in organic matter will start to generate oil, which is succeeded by 'cracking' of oil to generate gas.

Figure 6.4 Naturally occurring oil seep from Azerbaijan.

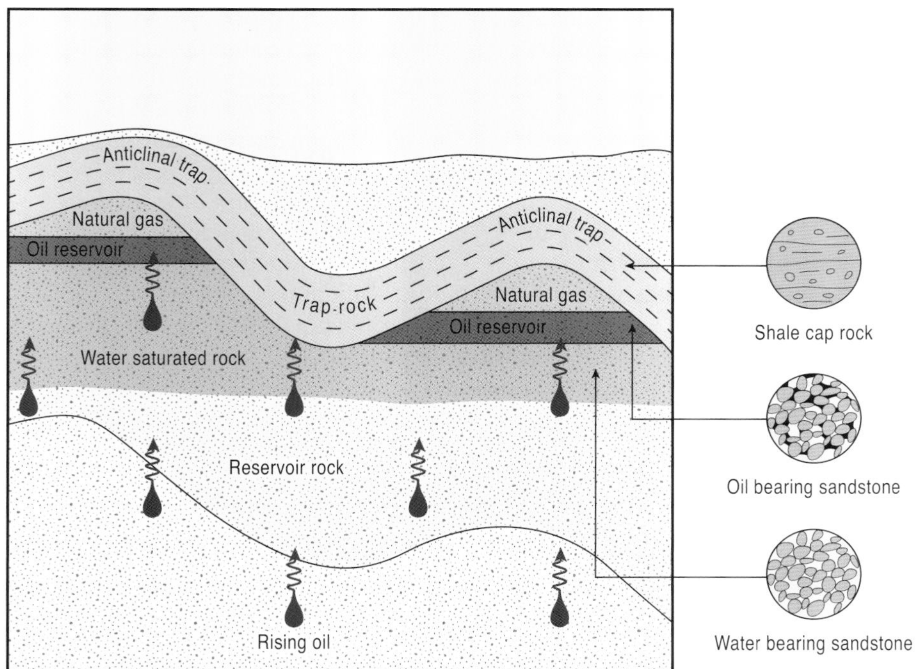

Figure 6.5 An example of a hydrocarbon trap. A trap is where a seal rock (e.g. shale or evaporite) overlies a reservoir rock (e.g. sandstone or limestone) and stops further migration of the oil and gas. In this example oil and gas are generated from a source rock and rise through overlying permeable sediments to accumulate at the crest of the fold. Oil and gas will displace water away from the crest, and over time allow economic volumes of oil and gas to accumulate.

up to the surface. If the oil is under pressure, it may move up to the surface freely, but more commonly the recovery of the oil and gas is enhanced through adding chemicals to reduce viscosity or by flushing with steam. The many advancements in oil and gas recovery from reservoirs still only produces about 40–50% of the oil and gas available in the sedimentary reservoir rocks.

Shale gas

Conventional gas reservoirs are created when natural gas migrates toward the Earth's surface from an organic-rich source rock into a highly permeable reservoir rock, where an overlying layer of impermeable rock traps the gas. In contrast, shale gas resources form within the organic-rich shale source rock. The low permeability of the shale greatly inhibits the gas from migrating to more permeable reservoir rocks (Table 6.1).

Over the past decade, the combination of horizontal drilling and hydraulic fracturing (informally known as fracking) has allowed access to the large volumes of shale gas that were previously uneconomical to produce (Fig. 6.6). The production of natural gas from shale formations has rejuvenated the natural gas industry in the United States. Fracking is a technique in which water, various chemicals, and loose sand, preferably with rounded grains, are pumped into the well to unlock the hydrocarbons trapped in shale formations by opening cracks or fractures in the rock and allowing natural gas to flow from the shale into the well. The sand grains help to prop open the fractures and allow the gas to flow more freely (Fig. 6.6). When used in conjunction with horizontal drilling, fracking enables gas producers to extract shale gas at reasonable cost. Without these techniques, natural gas would not flow to the well rapidly enough for commercial quantities to be produced from shale reservoirs.

Coal

Worldwide, coal is the most abundant and most widely distributed of the fossil fuels. At current rates of production, the reserves could last for as long as another two centuries (Fig. 6.7). It is worth stating that although coal is a fossil fuel like oil and gas, in contrast it forms from plant material (wood, stems and leaves) that once grew in a swampy environment similar to the wetland and rainforests of modern semitropical to tropical coastal areas (Fig. 6.8).

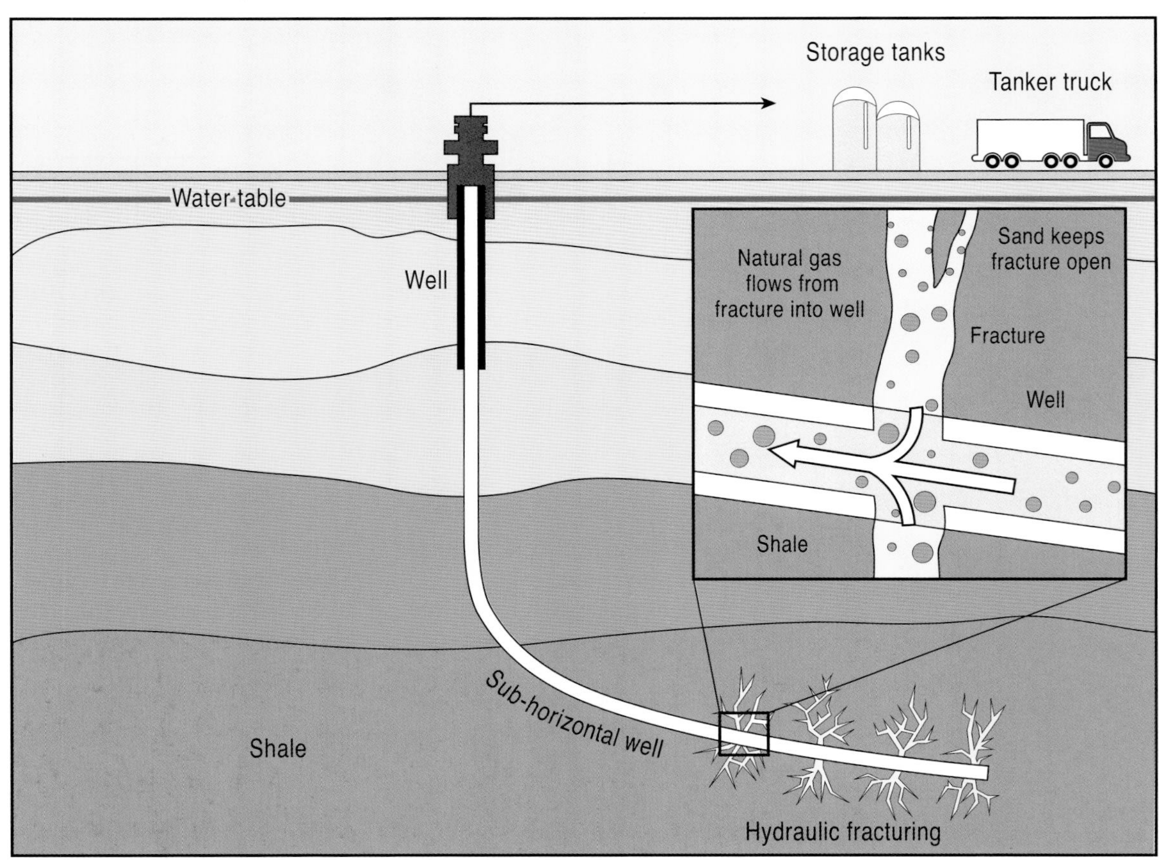

Figure 6.6 Hydraulic fracturing, commonly known as hydrofracking or simply fracking, involves pumping water mixed with sand grains and a variety of chemicals into horizontal underground gas wells at high pressure to form fissures in the shale to release natural gas stored in the pore spaces within the sedimentary rock.

Figure 6.7 Open pit mining from Fife, Scotland. © Leigh Sharpe and the Coal Authority.

The most extensive deposits of coal globally occur in sedimentary strata of Carboniferous age (298–358 million years ago). The abundance of coal at this time is due to two geological consequences; firstly, two large ice sheets at the southern pole locked up large amounts of water as ice. This caused a lowering of sea level and an increase in ideal environments for coal formation; and secondly, the past position of the continents straddled the Equator, providing warm tropical climates in which the Carboniferous vegetation flourished. The Carboniferous coal deposits contain many extinct species of giant tree ferns, giant horsetails and conifers.

The formation of coal is very much dependent on the sedimentary environment. Vegetation must be buried in oxygen-poor stagnant waters and incorporated into the sedimentary strata. Compaction and partial decay of the vegetation transforms into peat. To transform peat into coal requires the peat to be deeply buried. With increasing burial peat will transform into a dark-brown coal called lignite. As lignite is further compressed and heated to 100–200°C it turns into a dull black bituminous coal. This process drives out the water and volatile

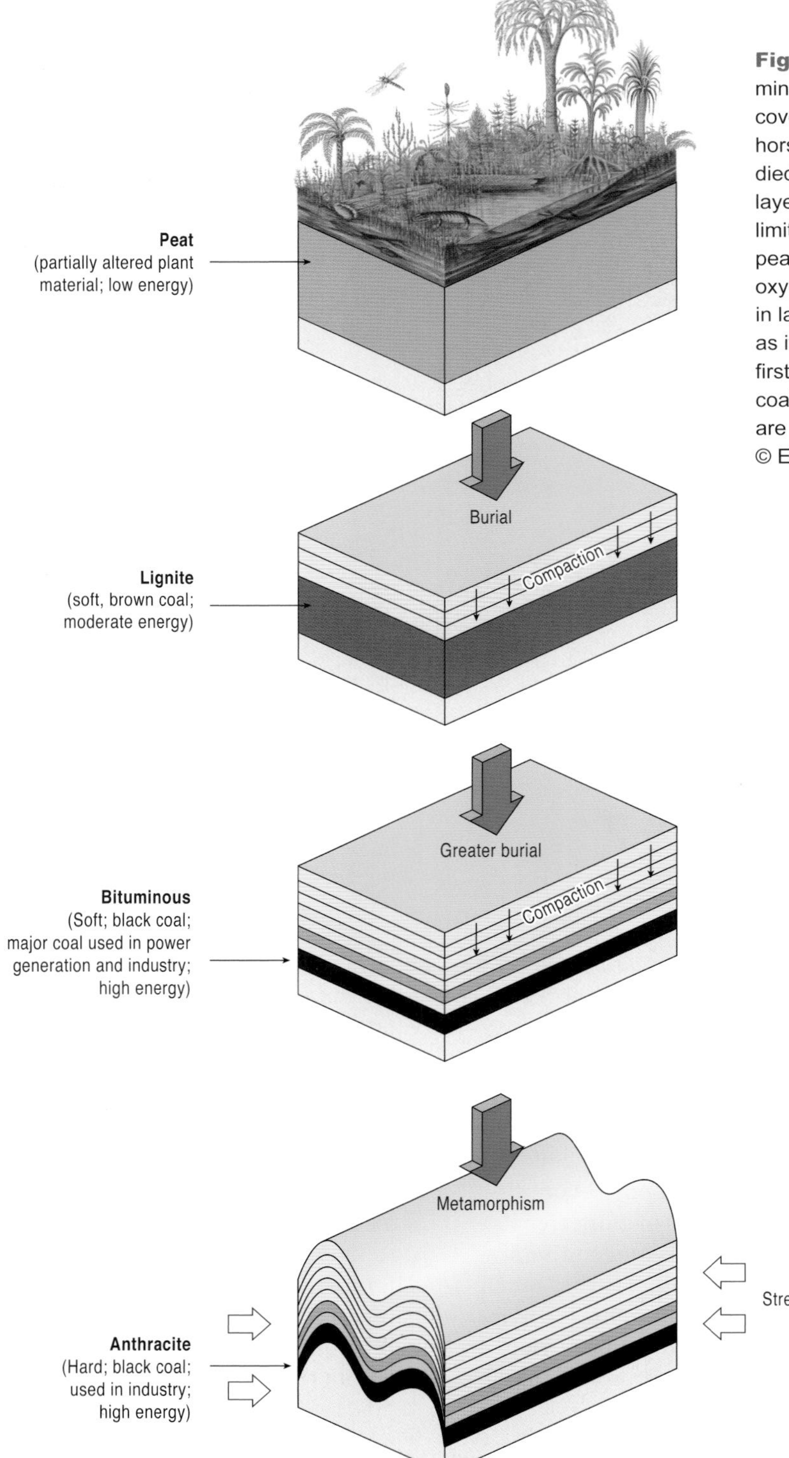

Figure 6.8 Coal is a black or brownish-black combustible mineral formed about 300 million years ago when the earth was covered by swampy forests of scale trees (lycopods) giant ferns, horsetails, and club mosses. Layer upon layer of these plants died and were compressed and then covered with soil. As the layers were successively covered their access to the air was limited and this stopped the full decomposition process, creating peat. Over the years heat and pressure worked to force out oxygen and hydrogen, leaving carbon-rich deposits, called coal, in layers known as seams. The carbon content of the coal rises as it is compressed further and the moisture content falls. The first type of coal to form is lignite, followed by sub-bituminous coal, bituminous coal and lastly anthracite. These grades of coal are known as ranks. (Carboniferous coal swamp reconstruction © Elizabeth Pickett.)

gases, concentrating the carbon. At still higher temperatures between 200 and 300°C, the bituminous coal will transform into shiny black **anthracite** with the highest carbon content of all coal types (Figs 6.8 and 2.9).

Anthracite is the highest rank of coal and the most highly prized by coal-miners (Fig. 2.9). Often called 'black diamond', it contains between 80 and 95% carbon and very low sulphur and nitrogen levels compared to lower ranking coals. It also burns at the highest temperature of all coals and produces the best heat output with little ash resulting.

Sedimentology will continue to play an important role for our future energy resources, in understanding how to extract the last drops of oil from a sandstone, working towards new ways of obtaining energy resources from sedimentary rocks, and considering how coal can be exploited in more energy-efficient ways.

Sedimentary mineral deposits

Chemical precipitation of metal ore minerals in sedimentary environments is the most common occurrence of iron, manganese and some copper ores. Sedimentary iron ore deposits are among the world's great mineral treasures. The most common recognized are **Banded Iron Formations (BIF)**, so called because of their alternating layers of grey iron oxide minerals (haematite, magnetite and siderite) and bright red chert (silica) (Fig. 6.9). BIF deposits were only formed between 2.5 and 1.8 billion years ago and provide significant information about the early atmosphere of the earth. This would suggest that following 1.8 billion years ago the ocean no longer contained abundant quantities of dissolved iron in seawater from

Figure 6.9 Precambrian age Banded Iron Formation (BIF) from the Hamersley Ranges of the Pilbara region in north-western Western Australia. © Graeme Churchard – Wikimedia common links.

which the iron minerals in BIF formed. Sedimentologists have hypothesized that the decrease in dissolved iron in seawater may have resulted from the onset of an oxygen-rich atmosphere, and iron cannot readily dissolve in oxygen-rich waters. BIFs are an important commercial source of iron ore from the Pilbara region of Western Australia (Fig. 6.9) and the Animikie Group in Minnesota, USA.

Placer deposits are also sedimentary in origin. They are found in rivers where the moving water has mechanically concentrated valuable minerals formed by gravity separation during the sedimentary processes. Placer minerals must be both dense and resistant to weathering processes. To accumulate in placers, mineral particles must be significantly denser than quartz (whose specific gravity is 2.65), as quartz is usually the largest component of sand or gravel found in most rivers. The action of the river concentrates the heavy minerals where the flow velocity is high enough to carry away lighter material, but leaves behind the heavy minerals. Such places include river bars on the inside of meander bends, plunge pools below waterfalls, and depressions on a riverbed (Fig. 6.10). Heavy minerals concentrated in this manner include gold dust and nuggets, native platinum and heavy oxides of tin and titanium. Furthermore, diamonds and other gemstones are often found in placers (Fig. 6.10).

Figure 6.10 Photograph of alluvial diamond mining along the Orange River, South Africa. Alluvial or placer diamond deposits occur where diamonds have been eroded from their primary source (kimberlites) over millions of years and transported by streams and rivers where they can become concentrated in sediments associated in a river bed. The sediments associated with the Orange River of South Africa are particularly plentiful in diamonds. Alluvial diamond mining can involve sucking the gravel and ancient fluvial sediments from the base of the river, where diamonds frequently accumulate, and the sediments then being taken to be washed and screened for diamonds.

Sedimentary rocks used in your house!

The society we all live in needs a wide variety of geological materials on a daily basis. Many of these tend to be of sedimentary origin. From the quarried sedimentary rocks for road stone and building, the chemicals for fertilizers, the gypsum for plaster board and building products, the quartz sand for glass making, clay for porcelain and ceramics, the products are endless (Fig. 6.11). Many of these sedimentary rocks do not have the glamour of other resources found in sediments, and their low unit price and high demand means they are usually locally sourced. The way we use sedimentary rocks is often best considered in our houses. The concrete foundations are made from limestone mixed with sandstone and mudstone in the correct proportions. The bricks originate from baking clay. The interior walls of modern houses are made from gypsum plasterboard. This comes from a slurry of water with the sedimentary mineral gypsum ($CaSO_4\ 2H_2O$) sandwiched between sheets of paper. The glass in the windows consists largely of silica formed by first melting and then freezing very pure quartz sand from a beach deposit or a sandstone formation. Without the many sedimentary resources of the Earth, modern society would grind to a halt.

Figure 6.11 An operational limestone quarry.

Glossary

A

arkose [14]: is a sandstone containing at least 25% feldspar. Arkose is generally formed from the weathering of feldspar-rich igneous or metamorphic rocks, most commonly granites which are primarily composed of quartz and feldspar.

anthracite [76]: a type of coal characterized by the high proportion of carbon and low content of volatiles. Anthracite is the highest grade of coal.

B

banded iron formations (BIFs) [77]: a distinctive sedimentary rock mostly confined to the Precambrian. Usually consists of repeating thin layers (a few mm to a few cm in thickness) of silver to black iron oxides, either magnetite (Fe_3O_4) or haematite (Fe_2O_3), alternating with bands of iron-poor shales and cherts, often red in colour.

bed [23]: a layer of sedimentary rock representing a single episode of sediment deposition, or a period of intermittent deposition of the same type of sediment, and bounded above and below by surfaces representing interruptions of sedimentation.

bedforms [30]: are irregularities or geometries formed on the surface of a bed of sand due to interaction between the flow (in water or air) and the grains of sand. Examples of bedforms include ripples and dunes.

biofacies [4]: a rock unit which contains an assemblage of fossils characteristic of a particular environment.

biogenic sedimentary rocks [13]: are sedimentary rocks derived from the skeletal remains and soft organic material of pre-existing organisms. Examples include most types of limestone, coal and some cherts.

biostratigraphy [61]: the branch of stratigraphy that utilizes fossils and fossil assemblages to date rock units.

bioturbation [34]: is the mixing and reworking of sediment by animals and plants. Patterns of bioturbation may be preserved in sedimentary rocks, and the study of such patterns is known as ichnology.

black shale [18]: these are dark-coloured mudrocks that are rich in organic matter (3–10% organic carbon). The organic matter found in black shales tends to comprise a high proportion of complex organic molecules derived from plankton which, when buried deeply enough (usually to depths of 3–4 km), break down to form oil shale; black shales are therefore often petroleum source rocks.

body fossils [36]: are the remains of the now fossilized organisms. Normally only the hard skeleton is preserved (shell or bone) and the soft tissue (skin, muscle, etc.) is lost. Animals that lack any skeleton or shell are very rarely fossilized. Examples of body fossils include ammonites, bivalves, trilobites and dinosaur bones.

Bouma sequence [57]: specifically describes the ideal vertical succession of structures deposited by low-density (i.e. low sand concentration, fine-grained) turbidity currents. The Bouma sequence is divided into five distinct layers labelled A through to E, with A being at the bottom and E at the top. Each layer described by Bouma has a specific set of sedimentary structures and a specific lithology, with the layers overall getting finer-grained from bottom to top. Most turbidites found in nature have incomplete sequences. Bouma describes the ideal sequence where all layers are present.

breccia [13]: coarse sedimentary rock consisting of angular to subangular fragments. Particular types of breccia can be recognized, including slump breccia, consisting of broken and brecciated beds derived from downslope slumps; collapse breccia, resulting from the dissolution of evaporates and the collapse of the overlying strata; and fault breccia, closely associated with fault zones and fault planes.

C

carbonate [9]: rocks containing calcite and /or dolomite.

cementation [13]: is the phase of lithification in which a cement precipitates from pore fluids in a sediment during diagenesis. Cementation is an important diagenetic process, and frequently during burial an early cement is replaced due to changing pore-water chemistry. Calcite, quartz and iron oxides are common types of cements.

chemogenic sedimentary rocks [13]: are sedimentary rocks that form from the direct precipitation of crystalline particulates from saturated and supersaturated solutions. Examples include evaporites, ironstones and some limestones (e.g. oolites).

clastic [9]: resulting from the disintegration of older rocks (sedimentary, metamorphic and/or igneous).

clastic sedimentary rocks [13]: are sedimentary rocks composed mainly of detrital grains derived from the weathering and erosion of pre-existing rocks. They include conglomerates, sandstones and mudrocks.

coal [17]: a sedimentary rock, used as a fuel, consisting mainly of plant remains in various stages of decay, ranging from lignite (organic- and volatile-rich) to anthracite (high carbon content).

compaction [13]: includes both the physical and chemical processes caused by the mass of the overlying sediment, which cause water to be squeezed out and grains to become more closely packed.

conglomerate [13]: a clastic sedimentary rock composed mainly of rounded clasts (rock fragments) larger than 2 mm in diameter.

coprolite [61]: fossilized faeces. The study of coprolites of fish, dinosaurs and marine reptiles can reveal useful information about diet.

cross-stratification [31]: stratification inclined to the original horizontal surface on which the sediment was deposited. The direction the cross-stratification is dipping indicates the palaeocurrent direction, the direction of sediment transport. Cross-stratification is most commonly produced by the migration of ripples and dunes. Also known as cross-bedding.

Current Ripples [30]: a depositional sedimentary structure formed by the movement of water. Usually asymmetric in profile and fairly straight crests in plan view. They commonly are about 2–5 cm in height and with wavelengths of up to ~40 cm.

D

depositional unit [6]: is a general term used to describe a distinct package of sediments, which may be a single bed or several beds, but with lower and upper confining boundaries. Refers to a single depositional environment.

diagenesis [20]: is the process by which sediments are lithified into sedimentary rocks and represents the sum of physical and chemical changes that occur during burial. Diagenesis involves physical compaction of components due to pressure increase on burial, the precipitation of mineral cements from pore fluids and phase transformations of mineral components. Diagenesis will continue with increasing pressures and temperatures until the onset of metamorphism.

dolomite [16]: is a carbonate rock containing more than 50% of the mineral dolomite ($CaMg(CO_3)_2$. Dolomite usually forms due to the replacement of calcite during diagenesis in a process known as dolomitization. This process usually erases most of the pre-existing texture of a limestone, producing a crystalline rock.

draas [34]: large-scale or mega-**dune** formed by the regional wind pattern. Often complex structures with dunes migrating over the top of the structures.

dune [31]: is a sandy sedimentary bedform deposited by wind or water and of similar geometry to ripples, but are much larger bedforms, with subaqueous dunes having a wavelength of >0.6 m and a height of >0.4 m. In comparison, subaerial or aeolian dunes are much larger again and range from 3 m to 600 m in wavelength and between 0.1 m and 100 m in height. Many aeolian dune sandstones are quartz arenites with rounded grains, which are frequently frosted due to grain abrasion.

E

epicontinental sea [9]: a shallow sea overlying a continent.

evaporites [18]: are chemogenic sediments that have been precipitated from water following the concentration of dissolved salts by evaporation. This can take place in both marine and non-marine (lake) waters. The principal evaporite minerals are gypsum, anhydrite and halite. Evaporites are an important economic mineral for manufacturing fertilizers and used widely in the building industry (e.g. plaster board manufacture).

F

facies [3]: a group of sedimentary rocks and primary structures indicative of a given depositional environment.

facies association [4]: a group of sedimentary facies used to define a particular sedimentary environment. For example, all the facies found in a deltaic environment may be grouped together to define a deltaic facies association.

feldspar [14]: a group of rock-forming silicate minerals that make up as much as 60% of the Earth's crust, e.g. plagioclase and K-feldspar. A common mineral component of many sandstones and especially arkoses.

flute casts [27]: are scours dug into soft, fine-grained sediment which typically get filled by an overlying sediment type (hence the name 'cast'). Found most commonly associated with turbidite sequences.

fossil Lagerstätten [64]: an example of well-preserved fossil material, either in great quantity, or exhibiting exceptional preservation.

G

geochronology [64]: the science of dating geological events in years.

geopetal structures [23]: a sedimentary fabric that records the way-up of the beds at the time of deposition. Geopetal structures are commonly found in cavity fills within limestones, where the lower part of the cavity has been filled with sediment and the upper part filled with a later cement.

greywacke [14]: is a sandstone containing more than 15% fine-grained matrix. Many greywackes form as part of turbidity current deposits and are poorly sorted sandstones.

groundwater [68]: water present within unconsolidated or permeable rock (often sedimentary rocks) beneath the water table.

H

hummocky cross-stratification (HCS) [34]: is a sedimentary structure found in sandstones that is characterized by cross-laminations with both concave and convex-upward forms. HCS is thought to form under a combination of unidirectional and oscillatory flow that is generated by relatively large storm waves in the ocean. They are usually characterized by long wavelengths (1–5 m), but with low heights (tens of centimetres).

I

ichnofacies [36]: a term used for a group of trace fossils found in a particular bed that indicate the nature of the geological setting and biological activity of the organisms that lived there. Ichnofacies has become an important tool for the sedimentologist to interpret ancient environments.

ichnology [36]: the science of the study of trace fossils as found in sedimentary rocks.

K

kerogen [70]: the naturally occurring solid, insoluble organic matter that occurs in source rocks (mudrocks and black shales) and can yield oil upon heating. Typical organic constituents of kerogen are algae and woody plant material.

L

lahar [19]: a type of mudflow or debris flow composed of a slurry of volcanic particles and water. The material flows down the flanks of a volcano, typically along a river valley. Lahars are extremely destructive.

limestone [15]: a sedimentary rock composed predominantly of calcite. Limestones comprise around 10% of sedimentary rocks.

litharenite [14]: are sandstones where the rock fragment content exceeds 25% and is greater than feldspar. Litharenites are immature sandstones, reflecting upon the types of rock fragments present. Many fluvial and deltaic sandstones are litharenites.

lithification [13]: the process of forming solid rock from sediment through the process of compaction and cementation.

lithofacies [4]: a distinctive group of characteristics within a rock unit, defined on the basis of the lithology (e.g. grain size and mineral content).

lithosphere [9]: the relatively rigid, outer 100 to150 km thick layer of the Earth, including the crust and uppermost part of the mantle. It is the lithosphere that is broken into tectonic plates.

M

mudrocks [14]: a sedimentary rock dominated by clay grade grains (<0.004 mm). The majority of grains in mudrocks are clay minerals such as smectite, illite and kaolinite.

N

Nicolaus Steno [4]: a Danish Catholic bishop and scientist who is credited with defining the basic principles of stratigraphy.

O

oil shale [18]: shale containing kerogen.

oil window [70]: the narrow range of temperatures under which oil and gas can form from an organic-rich source rock.

ooids, oolite [15, 16]: are small (typically 0.2-0.5 mm in diameter), spherical–subspherical grains, consisting of one or more regular concentric lamellae (coatings) of calcium carbonate. Ooids usually form on the sea floor, most commonly in shallow agitated tropical seas. Sediment composed of ooids is referred to as an oolite.

organic sedimentary rocks [13]: are sedimentary rocks that consist of carbon-rich relicts of plants and other organisms. The most common types are coal, lignite and oil shale.

P

placer deposits [78]: a sedimentary accumulation of economic minerals (ore) formed by sedimentary processes, e.g. in a river bed; heavy minerals such as gold and platinum often form such deposits.

principle of fossil succession [73]: in a stratigraphic sequence, different species of fossil organisms appear in a definite order (fossil succession) that can be recognized over wide horizontal distances. The fossil content, together with the **law of superposition**, helps to determine the time sequence in which sedimentary rocks were deposited.

principle of superposition [6 (Fig. 1.7a), 23]: sedimentary layers are deposited in a time sequence, with the oldest on the bottom and the youngest on the top.

Prokaryota [62]: primitive single-celled organisms that lack a defined nucleus or additional organelles.

pyroclastic fall [19]: pyroclastic fall deposits are those that have travelled through the air as some kind of projectile during a volcanic eruption (e.g. volcanic ash, volcanic bombs).

pyroclastic flow [19]: is a fast-moving current of hot gas and volcanic rock, which can reach several 100 km/h. They tend to hug the ground and travel downhill, or spread laterally under gravity. The speed of the flows depends on the density of the current, the volcanic output rate and the gradient of the slope.

pyroclastic surge [19]: is a current of turbulent gas and rock fragments which is ejected during some volcanic eruptions. It is similar to a pyroclastic flow but it has a lower density and contains a much higher ratio of gas to rock, which makes it more turbulent and allows it to rise over ridges and hills rather than always travelling downhill, as pyroclastic flows do.

Q

quartz [13]: an abundant, very hard mineral composed of silicon dioxide (SiO_2). It may form well-shaped hexagonal crystals and form in a variety of colours (e.g. rose quartz and amethyst). It is an important constituent of many sedimentary rocks, in particular sandstones and mudrocks.

quartz arenite [14]: is a sandstone, composed of 95% or more quartz grains, and is the most compositionally mature of all sandstones. Cements tend to be quartz, but calcite can occur. In many cases quartz arenites are the products of extended periods of sediment reworking so that all grains except for quartz have broken down.

R

regression [4]: the seaward migration of the shoreline caused by the lowering of sea level.

reservoir rock [70]: usually a sedimentary rock (e.g. sandstones and limestones) which contains hydrocarbons (oil and gas) that have migrated from a source rock (e.g. black shale).

rock cycle [11]: the succession of events that results in the transformation of Earth materials from one rock type to another and so on.

rounding [2, 12]: is the presence or absence of corners and sharp edges on grains. Grains with many edges are angular. Grains lacking edges are rounded. Most natural aeolian sand grains tend to be sub-rounded to rounded due to the strongly abrasive action of the wind currents.

S

sandstone [13]: also known as an arenite, is a clastic sedimentary rock comprised mainly of sand grade particles between 0.0625 mm and 2 mm. Sandstones are subdivided on the basis of their grain types. Grains can be quartz, feldspar or rock fragments (lithic). Sandstone types include quartz arenite, litharenite, arkose, and those with more than 15% fine-grained matrix are known as greywackes.

scour marks [27]: are a sedimentary structure produced as the result of erosion of a sediment surface by the current flowing over it. The soft but cohesive sediment surface, generally mud, is sculptured and reshaped by the scouring action of the current.

sediment [1]: an accumulation of loose mineral grains of different sizes, from boulders to mud, that are not cemented together and usually transported.

sedimentary basin [7]: a depression, created as the consequence of subsidence, that fills with sediment.

sedimentary structures [23]: any

structure in a sedimentary rock formed at or shortly after deposition (e.g. cross-bedding, flute marks).

sedimentary rock [1]: rock that forms either by the cementing of grains broken off pre-existing rock or by the precipitation of mineral crystals out of water (usually seawater) at or near the Earth's surface.

sedimentary sequences [7]: a group of sedimentary units bounded on top and bottom by regional unconformities.

sedimentology [1]: the study of modern and ancient sediments and the processes that result in their deposition.

septarian nodules [39]: are diagenetic concretions containing angular cavities or cracks, which are called septaria.

sequence boundaries [7]: are significant erosional unconformities as used in sequence stratigraphy. These boundaries are the product of a fall in sea level that erodes the exposed sediment surface of the earlier sequence or sequences.

stratigraphy [1]: the branch of geology dealing with rock successions and geological history.

sequence stratigraphy [6]: the study of sedimentary rock relationships within a time-stratigraphic framework of related facies bounded by widespread unconformities.

sorting [1, 12]: the process whereby clastic sediments are separated into groups of different grain size by the action of water and/or wind.

spreite [38 (Fig. 3.19)]: near tabular, curved laminae that are characteristic of certain trace fossils. They are formed by invertebrate organisms tunnelling back and forth through sediment in search of food.

strata [4]: layers in sedimentary rocks.

subsidence [7]: the process by which vertical sinking of the Earth's surface occurs relative to a reference plane/ crustal feature.

T

tool marks [27]: are formed when objects carried by currents impact or scrape along the surface of a sedimentary unit. Tool marks occur as various types from bounce and prod markings to more continuous grooves and chevrons orientated parallel to the flow direction.

transgression [4]: the inland migration of the shoreline from a rise in sea level, resulting in flooding.

travertine [15]: is a form of limestone deposited especially at hot-water springs. Travertine is a distinctive finely crystalline limestone, often with concentric lamination and radial patterns of crystal growth. It may occur purely as an inorganic precipitate from geothermally heated alkaline waters, but can also be influenced by microbial activity.

turbidity current [14]: a water current generated by gravity-induced flow, carrying large quantities of sediment of varying coarseness in suspension. The sedimentary deposits are known as turbidites.

V

ventifact [46]: a pebble whose surface has been polished, etched, grooved, pitted or faceted by wind–driven sand in arid environments. They can be used to determine palaeo-wind directions.

volcanoclastic sedimentary rocks [13]: sedimentary rocks composed chiefly of grains and clasts derived from volcanic activity.

W

water table [68]: the boundary that separates regions saturated with groundwater from those unsaturated. The water table may vary due to seasonal changes such as precipitation and evapotranspiration.

wave ripples [32]: ripples characterized by a symmetrical profile and continuous straight crests. They are common in many shallow marine to intertidal and deltaic sediments, both sandstones and limestones.

way-up criteria [23]: criterion used to recognize if a sequence of sedimentary rocks are the correct way-up or have been overturned by subsequent deformation events.

William 'Strata' Smith [4]: William Smith, also known as 'Strata Smith', was a surveyor who drained marshes and built canals in England in the Industrial Revolution. He made the important connection between fossil occurrences and the layer of sedimentary rocks they were in, and used this to create the first geological map of England and Wales.

Further reading

The following books provide more detail about sedimentology and takes many of the topics covered in this book to a more advanced level:

Boggs, S., 2009. Petrology of sedimentary rocks, Cambridge University Press.

Collinson, J.D., Mountney, N.P., and Thompson, D.B. 2006. *Sedimentary Structures*, Terra Publishing, Third edition.

Leeder, M.R., 2011. *Sedimentology and sedimentary basins: From turbulence to tectonics*. Wiley-Blackwell Publishing, Second edition.

Nichols, G., 2009. *Sedimentology and stratigraphy*. Wiley-Blackwell Publishing, Second edition.

Stow, D.A.V., 2005. *Sedimentary rocks in the field: a colour guide*. Manson Publishing, London.

Tucker, M.E., 2001. *Sedimentary Petrology: an introduction to the origin of sedimentary rocks*. Wiley-Blackwell Publishing, Third edition

Tucker, M.E., 2011. *Sedimentary rocks in the field: a practical guide*. Wiley-Blackwell Publishing, Fourth edition.